T0327367

SUBSTATION AUTOMATION SYSTEMS

SUBSTATION AUTOMATION SYSTEMS

DESIGN AND IMPLEMENTATION

Evelio Padilla
Eleunion C.A., Caracas, Venezuela

This edition first published 2016

Registered Office
John Wiley & Sons, Ltd., The Atrium, Southern Gate, Chichester, West Sussex, PO19 8SQ, United Kingdom

For details of our global editorial offices, for customer services and for information about how to apply for permission to reuse the copyright material in this book please see our website at www.wiley.com.

Library of Congress Cataloging-in-Publication Data

Padilla, Evelio.
Substation automation systems : design and implementation / Evelio Padilla, Eleunion, Venezuela.
    pages cm
  Includes bibliographical references and index.
  ISBN 978-1-118-98720-9 (cloth)
1. Electric substations–Automatic control. I. Title.
  TK1751.P33 2016
  621.31′26–dc23

2015021965

A catalogue record for this book is available from the British Library.

Set in 10/12pt Times by SPi Global, Pondicherry, India

Printed in the UK

Contents

Preface

A number of technological changes have occurred in the substation environment over the last 30 years. Surge arresters built with metal oxide discs, circuit breakers isolated with SF6 gas, numerical protective relays and other novel products that appeared early in the 1980s, were quickly adopted without significant impact on substation design. A few years after, however, the incursion of digital technology caused a "jerk" in the field of substation secondary systems. While young system engineers with a limited knowledge in substation-related concepts have become engaged in development of the engineering process Substation Automation Systems (SASs) from the side of device manufacturers, experienced utilities personnel had to (and in some cases still need to) face up to many disconcerting and complex scenarios characterized by an unusual lexicon and a lot of abstract resources that are now being applied to define and implement control and monitoring functionalities in their substations.

This book intends to help both professional groups accomplish their responsibilities by giving them guidelines with respect to the scope and functions of SASs based on current technology, including requirements from Standard IEC 61850, as well useful details for dealing with various stages needed for SAS project development.

The material is organized into 19 chapters; Chapter 1 providing a brief review on how SAS has recently evolved, Chapter 2 outlines the purpose of the SAS as an essential part of the substation, in Chapter 3 the effects of Standard IEC 61850 on different stages of SAS projects are presented, Chapter 4 illustrates constructive and functional features of equipment that make up the primary power circuit, Chapter 5 introduces the characteristics of Intelligent Electronic Devices (IEDs) used for control and monitoring and describe briefly certain phenomenon able to affect in detrimental way the physical/functional integrity of such devices, Chapter 6 provides an overview of how the features and functions of devices installed into the main control house are used for controlling and monitoring the substation as a whole, Chapter 7 contains different SAS functionalities including switching commands and constraints like interlocking and blocking conditions, Chapter 8 shows the set of signals coming from different substation components that need to be managed by the SAS, Chapter 9 suggests how the SAS ought to be engineered, Chapter 10 covers the theory and practical principles that support a typical implementation needed for the substation control and monitoring from a remote master station, Chapter 11 describes a lot of items that may characterize the SAS structure including options for the network topology further to quality requirements and cyber-security considerations, Chapter 12 contains recommendations regarding the tests to carry out on SAS components, Chapter 13 may serve as a baseline for programming and checking results of Factory Acceptance Tests (FATs) performed on representative SAS segments, Chapter 14 covers site testing scope and strategies, Chapter 15 proposes scope and sequence of training programs addressed to utilities personnel, Chapter 16 outlines how to deal with SAS projects, Chapter 17 offers a number of tips useful to help in getting timely acceptable SAS components and functionalities, Chapter 18 summarizes resources to be used and methodology to be followed for the engineering process according to Standard IEC 61850, and finally, Chapter 19 forecasts where control and monitoring technologies may go in the future.

In summary, the book intends to serve the practical needs of different participants in SAS projects with respect to technical matter and also from the management perspective.

Evelio Padilla

Acknowledgments

I would like to gratefully acknowledge several people for their valuable help on this book.

Firstly to all Wiley staff including Laura Bell, Assistant Editor, who was my initial contact with the company (and subtly followed with the publishing idea); Peter Mitchell, Publisher, Electrical Engineering, who achieved the tremendous goal of getting approval for the book project; Ella Mitchell, Associate Commissioning Editor, ever enthusiast in charge of extensive previous details and formal arrangements for the book project; Richard Davis, initial Project Editor, who gave me the guidelines related to the process for writing the book; Prachi Sinha Sahay, temporary Project Editor, who with patience and wisdom dealt with the completed manuscript, and Liz Wingett, the Project Editor who masterly cooperated in the process of adding value to the entire manuscript.

Thanks to Professors Nelson Bacalao and Greg Woodworth of Siemens Energy, who, as reviewers of the book project, found it to be feasible and contributional to the power industry.

My daughter, Jessenia, and her husband, Vinicio, provided me with precious and constant support and encouragement during the preparation of the manuscript.

My partner, Maria, was generous in her understanding and patience during the writing process.

I am also grateful for the training and support received from my employer EDELCA (currently CORPOELEC), as well as Carabobo University for my higher degree education.

CORPOELEC kindly gave their permission to include all the photos that appear in the book.

Yunio Leal, Julio Aponte and Zurima Alfonzo were outstanding collaborators during my working stay in EDELCA.

My appreciation is also expressed to my parents, Juanpa and Clara Rosa, who supported my early education; to my brother, Elias, and my sister, Argelia, for always identifying with my professional career; my sons, Armando and Alejandro, permanent inspiration sources, and mathematician, Professor Daniel Labarca, who inspired me to become an engineering professional.

Finally, the biggest thanks to God; because without His guidance, nothing is possible.

Evelio Padilla

List of Abbreviations

AC	Alternating voltage system
A/D	Analog/digital
APT	Auxiliary power transformer
AV	Analog value
BB	Busbar
BC	Bay controller
BC-AS	Bay controller for auxiliary system
BF	Breaker failure
BI	Binary input
BIL	Basic impulse level
BO	Binary output
BPD	Bushing potential device
CB	Circuit breaker
CIGRE	International Council on Large Electric Systems (Conseil International des Grands Réseaux Électriques)
CPU	Central processing unit
CT	Current transformer
DB	Database
DC	Direct voltage system
DG	Diesel generator
DI	Disconnector
DR	Disturbance recorder
DNP	Distributed network protocol
EMC	Electromagnetic compatibility
EMI	Electromagnetic interference
ES	Earthing switch
GOOSE	Generic object oriented substation event
GPS	Global positioning system
HMI	Human machine interface

HV	High voltage
HW	Hardware
IEC	International Electrotechnical Commission
IEEE	Institute of Electrical and Electronics Engineers
IED	Intelligent electronic device
IT	Information technology
I/O	Input/output
LAN	Local area network
LCD	Local control display
LV	Low voltage
MCB	Mini circuit breaker
MMS	Manufacturing message specification
MTTF	Mean time to failure
MU	Merging unit
MV	Medium voltage
MVA	Mega-volt ampere
NCC	Network control center
OLTC	On-load tap changer
OPGW	Optical grounding wire
PB	Process bus
PC	Personal computer
PCG	Protocol converter gateway
PR	Protective relay
PT	Power transformer
RTU	Remote terminal unit
SAS	Substation automation system
SB	Station bus
SC	Station controller
SCL	Substation configuration description language
SF6	Sulfur hexafluoride
SLD	Single line diagram
SOE	Sequence of events
SV	Sampled values
SVC	Static var compensator
VT	Voltage transformer

1

Historical Evolution of Substation Automation Systems (SASs)

The key goal in the operation of electrical power systems is to maintain the energy balance between generation and demand in an economic manner. This often requires changes in system configuration to keep voltage and frequency parameters at acceptable pre-specified ranges; furthermore, configuration changes are needed for maintenance work at utility installations or for clearing faults due to short-circuit currents. Typical changes in system configurations include connection and disconnection of generators, power transformers, transmission lines, shunt reactors and static reactive power compensators. Therefore, such changes in system configuration are made through control facilities available in both generation stations and substations located along transmission and distribution systems (see a view of a substation in Figure 1.1).

Until a few decades ago, the control of electric substations was based on systems consisting of discrete electronic or electromechanical elements, where several functions were carried out separately by specific subsystems. Although those arrangements were reliable because the failure of a subsystem does not affect the performance of the rest of control facilities, it was also quite expensive, as they require a large investment in wiring, cubicles and civil engineering work. Back then, stations were controlled through a large mimic control board located in the main control house, as shown in Figure 1.2.

Sometimes, primary arrangements of substations were placed outside control cubicles lodged in dedicated relay rooms (Figure 1.3).

One of the most emblematic components of that age was the flag relay shown in Figure 1.4, which was the main way to display alarms for the attention of the substation operator.

In terms of civil engineering work, some substations were provided with large concrete channels where several kilometers of copper cables were run, as shown in Figure 1.5.

When microprocessor based substation control systems were originally developed, they were conceived as RTU-centric architecture, and later a distributed LAN architecture became the predominant technology. In more recent years, when control systems and other secondary systems began to incorporate new communication technologies and Intelligent Electronic

Substation Automation Systems: Design and Implementation, First Edition. Evelio Padilla.

Figure 1.1 View of a 765 kV electric substation. Source: © Corpoelec. Reproduced with permission of Corpoelec

Figure 1.2 Old mimic control board. Source: © Corpoelec. Reproduced with permission of Corpoelec

Figure 1.3 Substation primary arrangement shown outside control cubicles. Source: © Corpoelec. Reproduced with permission of Corpoelec

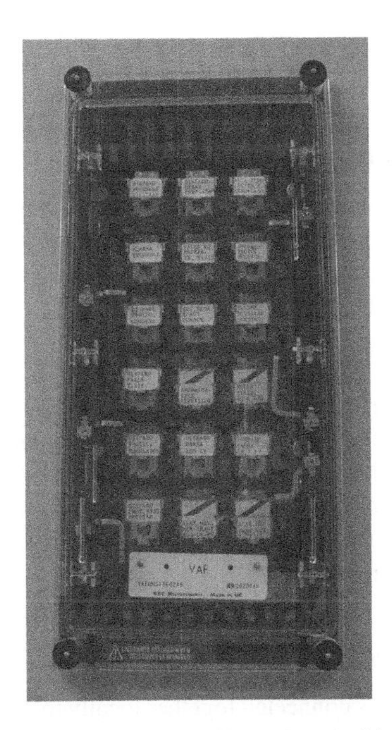

Figure 1.4 Flag relay. Source: © Corpoelec. Reproduced with permission of Corpoelec

Figure 1.5 Old cabling channels. Source: © Corpoelec. Reproduced with permission of Corpoelec

Devices (IEDs), the complete set of secondary facilities and functionalities was referred to as "Substation Automation Systems" (SASs).

1.1 Emerging Communication Technologies

Development of communication technologies represents an important step allowing SASs to be more and more versatile and increase functionality. The most influential new technologies applied in substations are described in the following sections.

1.1.1 Serial Communication

Serial communication is the process of sending data one bit at a time, over a single communication line. In contrast, parallel communication requires at least as many lines as there are bits in a word being transmitted. This kind of communication was widely used at the beginning of the digital technology incursion in substations; in particular for relay to relay connections through a RS-232 interface. In recent years, instead of serial communication, Ethernet connectivity is gaining a place.

1.1.2 Local Area Network

As a group of computers/devices connected together locally to communicate with one another and share resources, this solution was early dedicated to office environments and later introduced to industrial applications, including substations. The use of LANs in a substation is increasing, in particular the Ethernet LAN specified in Standard IEEE 802.3.

1.2 Intelligent Electronic Devices (IEDs)

Generally, this refers to any device provided with one or several microprocessors able to receive/send data to or from another element. The most common IEDs used in substations are the following types:

1.2.1 Functional Relays

Digital relays (sometimes called computer relays, numerical relays or microprocessor-based relays) are devices that accept inputs and process them using logical algorithms to develop outputs addressed to make decisions resulting in trip commands or alarm signals. Early on, this kind of relay was designed to replace existing electromechanical or electronic protective relays and some years later they were also extended for use in control and monitoring functions.

1.2.2 Integrated Digital Units

Integrated digital units (also called multifunctional relays) have been developed for improving the efficiency of the substation secondary system decreasing the total cost of the asset by adding, in one element, several functions such as protection, control, monitoring and communication. This kind of device is widely used in particular for medium voltage substations where required availability is not a critical aspect.

1.3 Networking Media

The physical structure of LANs compresses cabling segments and connectivity devices allow computers and other IEDs connected to the LAN to share data and communicate. In the past, these elements were made up of copper wires and standardized communication ports and interfaces. Nowadays, these networking media are made with the following resources:

1.3.1 Fiber-Optic Cables

The use of optical technology eliminates the need for thousands of copper wires in a substation and replaces them with a few fiber-optic cables, making savings derived from installation and maintenance work while at the same time increasing worker safety and power system reliability. The main technical advantages in using fiber-optic cables in substations include high immunity to electrical interference and generous bandwidth. Today, the industry offers standardized fiber systems compatible with IEC 61850 devices oriented at reducing the chance of mistakes and minimizing costs in testing and commissioning activities.

1.3.2 Network Switches

These components are required to network multiple devices in a LAN. Their main function is to forward data from one device to another on the same network. They do it in an efficient manner since data can be directed from one device to another without affecting other devices on the same network. The most popular network switch used today in substations is the Ethernet switch, with different features or functions.

1.4 Communication Standards

Standards development is currently like "the motor" for SAS evolution. Initially, the Standard IEC 61850 had solved the important paradigm of vendor dependence that was blocking the advances in digital SAS installation for some years. Now, the Standard IEEE 803.2 allows an increase in networking facilities and functionalities. Both standards represent the state of art of SAS design and implementation as we know today, bringing clear rules for hardware design trends by manufacturers and more confidence in SAS users worldwide.

1.4.1 IEC Standard 61850 (Communication Networks and Systems for Power Utility Automation)

This Standard is a collection of publications intending to satisfy existing and emerging needs of the power transmission industry keeping interoperability as the main goal (allowing IEDs provided by different vendors to exchange data and work together in an acceptable manner). The Standard is based on continuous research and studies carried out by prestigious institutions such as UCA, CIGRE and IEEE, as well as the IEC itself. The scope of the standard currently is mainly addressed at the following:

- Technically define communication methods and specify their quality attributes.
- Provide guidelines for SAS project management and network engineering.
- Give recommendations for SAS testing and commissioning.
- Establish procedures for communication between substations.
- Define methods for communication between substations and remote control facilities.
- Provide guidelines for wide area control and monitoring.

The IEC still continues to develop several new areas. Utilities are waiting for them.

1.4.2 IEEE Standard 802.3 (Ethernet)

This standard defines the communication protocol called the "Carrier Sense Multiple Access Collision Detect" (CSMA/CD), which works under the broadcasting principle of carrying all delivered messages to all IEDs connected to a LAN. Currently, the standard maintains leadership on a LAN substation since such protocols were adopted by the IEC 61850 Standard as their communication platform. IEEE is very active in introducing innovations as the basis for network protocols, leaving behind a long history of proprietary protocols.

All these technological changes now provide the opportunity to have comfortable solutions for SAS design at reasonable cost and with reasonable levels of risk, in such a way that modern SAS are equipped with clean and sophisticated control cubicles lodged in appropriate control rooms (see Figure 1.6), and are operated from the main control house by means of ergonomic control desks such as that shown in Figure 1.7. This allows substation owners to get a high performance system characterized by excellent availability and reliability.

Figure 1.6 Control cubicles of a modern SAS. Source: © Corpoelec. Reproduced with permission of Corpoelec

Figure 1.7 Operation desk of a modern SAS. Source: © Corpoelec. Reproduced with permission of Corpoelec

Further Reading

Andersson, L. and Wimmer, W. (October 2001) Some aspects of migration from present solutions to SA systems based on the communication Standard IEC 61850, *2nd International Conference: Integrated Protection, Control and Communication – Experience, Benefits and Trends,* New Delhi, India.

Brand, K.P. (2004) Design of IEC 61850 based substation automation systems according to customer requirements, Paper B5–103, CIGRE Session 2004.

CIGRE Working Group 35.04 (February 2001) Optical Fiber cable selection for electricity utilities *Electra Review* 194, 49–51.

CIGRE Working Group B5.07 (April 2004) The automation of new and existing substations: Why and how. *Electra Review* 215, 53–57.

CIGRE Working Group D2.13 (August 2005) Web technology for the utility environment. Communication networks for web-based and non web-based applications. *Electra Review* 221, 30–36.

IEC 61850, Communication networks and systems for power utility automation, (www.iec.ch).

IEEE 802.3, Standard for information technology – Telecommunications and information exchange between systems – Local and metropolitan area networks, Part 3: Carrier sense multiple accesses with Collision Detection (CSMA/CD) Access Method and Physical Layer Specifications.

Kruimer, B. (August 2003) Substation Automation – Historical Overview, *IEC Seminar*, Kema, Amsterdam.

Madren, F. (2004) *Ethernet in Power Utilities Substations – The Changing Role of Fiber Media*, GarrettCom, Inc.

Padilla, E. and Ceballos, L. (2006) Managing the transition to IEC 61850 standard on Substation Automation Systems, *China International Conference on Electricity Distribution.*

Singh, N. (February 2002) Substation control in the system control. *Electra Review* 200, 41–53.

Tangney, B. and O'Mahony, D. (1988) *Local Area Networks and their Applications*, Prentice Hall, Hertfordshire.

Wester, C. and Adamiak, M. (n.d.) *Practical Applications of Ethernet in Substations and Industrial Facilities*, Internal paper GE Digital Energy Multilin.

2

Main Functions of Substation Automation Systems

Physically, all electric substations are comprised of a group of high voltage apparatus whose individual sizes depend on substation operation voltages, which is called the *primary equipment*, as well as by a lot of low voltage smaller components, which as a whole are called the *secondary system*.

The group of high voltage apparatus comprises changing-state equipment or switchgear (circuit breakers, disconnectors and earthing-switches, see Figures 2.1, 2.2 and 2.3) used to maintain or to interrupt the energy flux from/to transmission lines or load feeders connected to the substation, instrument transformers (voltage transformers and current transformers) that reflect voltages and currents present at high voltage terminals of primary equipment, and also in most cases, power transformers to change voltage levels according to substation purpose (Figure 2.4).

The secondary system consists of a less visible set of artifacts, which include components and facilities needed by the power system operator to make changes in the power system configuration opening or closing switchgear, relays to protect power system segments from short-circuits, overloads and other dangerous conditions, internal power sources to serve all substation electricity requirements (Figures 2.5 and 2.6) and other different components used to support the substation performance in a safe and reliable manner.

Substation automation systems (SASs) are based on a lot of dedicated software stored in pieces of hardware that belong to a set of substation secondary components. In a simple approach, using more recent technology, SASs are composed mainly of three groups of devices plus two Local Area Networks integrated as shown in Figure 2.7. The process devices group includes analog/digital converters and actuator devices to make the transition between SAS and high voltage equipment. The interface devices group covers a set of Intelligent Electronic Devices (IEDs) that receive and process signals coming from high

Substation Automation Systems: Design and Implementation, First Edition. Evelio Padilla.

Figure 2.1 View of a 400 kV circuit breaker. Source: © Corpoelec. Reproduced with permission of Corpoelec

Figure 2.2 View of a 400 kV disconnector (in the closed position). Source: © Corpoelec. Reproduced with permission of Corpoelec

Figure 2.3 View of a 115 kV earthing switch. Source: © Corpoelec. Reproduced with permission of Corpoelec

Figure 2.4 View of a 765 kV power transformer. Source: © Corpoelec. Reproduced with permission of Corpoelec

Figure 2.5 View of a distribution cubicle. Source: © Corpoelec. Reproduced with permission of Corpoelec

Figure 2.6 View of a battery room. Source: © Corpoelec. Reproduced with permission of Corpoelec

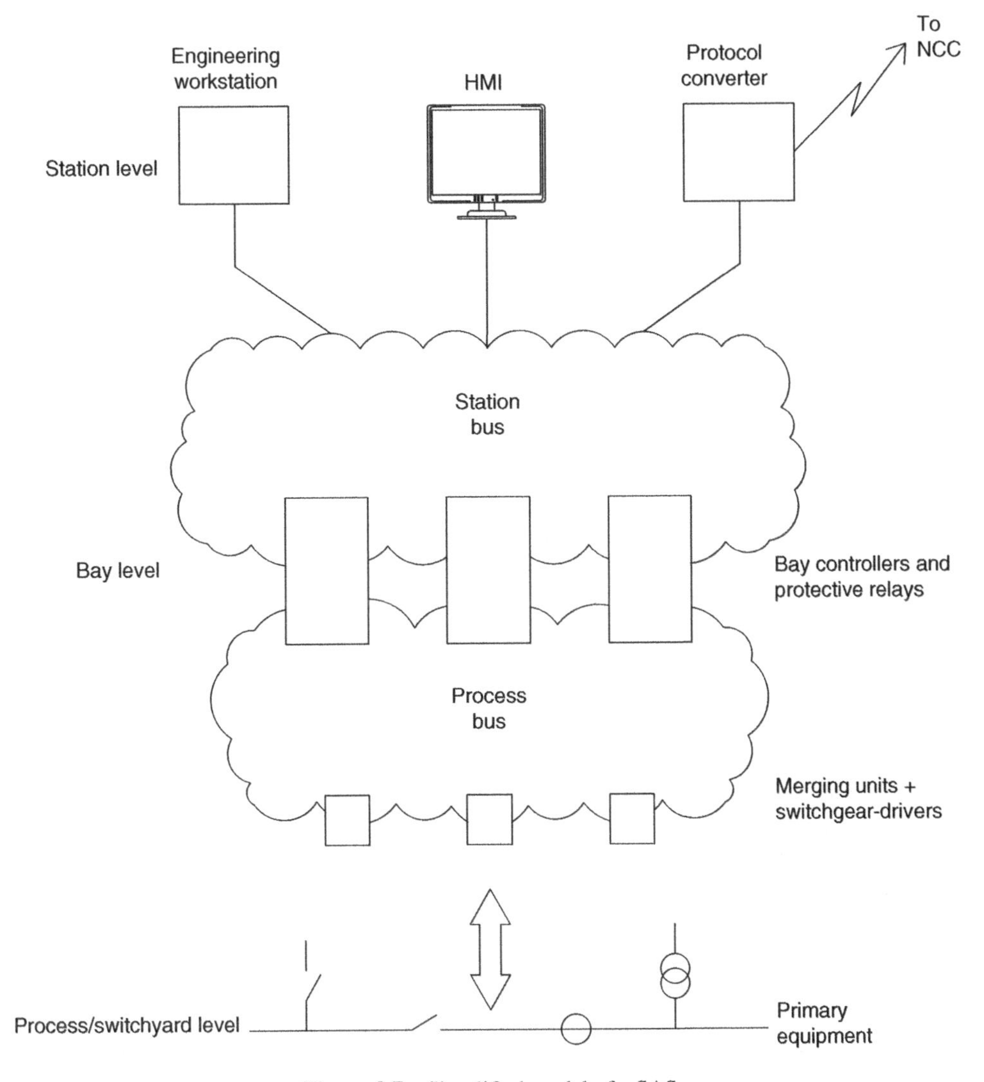

Figure 2.7 Simplified model of a SAS

voltage equipment. The application devices group includes all computers and other components required to run control functionalities and to communicate with internal and external subsystems.

The most important functions of a SAS are:

- Control:
  - Selecting, opening and closing circuit breakers and disconnectors.
  - Blocking and unblocking control commands.
  - Giving release information to circuit breakers and disconnectors for securing the opening and closing actions.

- Monitoring:
  - Showing substation configuration with position indication (open or closed) of circuit breakers and disconnectors based on signals coming from their own position contacts.
  - Acquiring and process data coming from power transformers and other primary equipment related to condition operation.
  - Displaying substation events including information regarding switchgear opening and closing actions due to any external cause, such as the activation/operation of a protective relay.
- Alarming:
  - Announcing to a substation operator all adverse conditions that may represent a risk to substation integrity.
  - Preventing trouble with SAS operation.
- Measurement:
  - Acquiring and showing current values of electrical or other relevant parameters.
  - Giving indications of energy flows through substation primary equipment and transmission lines.
- Setting and monitoring of protective relays:
  - Allowing changes on operating parameters of protective relays.
  - Giving alarm signals when any undesirable condition may affect the right relay performance.
- Control and monitoring of the auxiliary power system:
  - Displaying screens/drawings showing the configuration of the auxiliary power system.
  - Allowing selection and execution of control commands.
  - Driving automatic transfer switches.
  - Managing interlocking logics.
  - Supervising AC/DC power source conditions.
  - Giving alarm signals from abnormal conditions.
- Voltage regulation:
  - Monitoring actual voltage value on the power system.
  - Changing the position of the tap-changer of power transformers.
  - Giving alarms and signals.

In summary, SASs take care of all data acquisition process, control, monitoring and alarming functions associated with high voltage apparatus that belong to primary equipment, as well as similar functions associated with secondary substation systems. The control and monitoring information is presented to the operator through a graphical interface that shows overview diagrams, control means, alarms, measurement, trends and sequences of events displayed on user-friendly screens.

2.1 Control Function

To make changes in the primary arrangements that are required for power systems operation, the system operator, through the SAS, opens (or closes) circuit breakers and disconnectors installed at the substation. Switching operations can be made in a hierarchical order from different physical locations, the most common are:

- The switchyard itself (process/switchyard level)
- The local control room (bay level)

- The main control house (station level)
- The network control center (remote/NCC level).

The SAS displays screens with different features. At least one of them consists of a control dialog window that allows the substation operator to select the circuit breaker or disconnector to be open or closed. As a second step, the switching command is delivered.

When the switching command is for the closing operation of a circuit breaker, a previous check synchronization process takes place. Such process verifies voltage values on both sides of the selected circuit breaker, the voltage difference across it and phase-shift between voltages. The closing operation of a circuit breaker is permitted according to a choice of one or several of the following conditions:

- Voltage is only present in the substation busbar; that is, the feeder is dead.
- Voltage is only present in the feeder; that is, the substation busbar is dead.
- Voltages are present in both substation busbar and feeder and lie within permissible ranges for the conditions.

Other previous checking corresponds to the interlocking condition given by internal interlocks of the switchgear, interlocks at local bay level and substation level interlocking. These interlocking conditions cover, without exception, all relevant switchgear positions and switching operations at all control levels to ensure proper personnel safety and protection of the substation equipment.

In addition, all switching commands are allowed when no blocking condition is present. These blocking conditions may come from high voltage equipment (e.g., low gas pressure on the selected circuit breaker) and also from the SAS itself (e.g., disregard for the control hierarchy).

2.2 Monitoring Function

The SAS displays a group of mimic diagrams showing sections of the substation with color-coded symbols and position information of primary equipment and secondary relevant components. Provision is made to display lists of events occurring in the substation and related to transmission lines, as well as lists of messages and alarms.

Single-line diagrams show the status of all the switchgear in the substation in a simplified form so that the substation operator has a quick overview with regard to:

- Which feeder is connected to which busbar
- The actual busbar configuration
- Those feeders that are generating signals.

The position of the switchgear is represented by effective visual means, for example:

- Switchgear in the closed position = Steady light in symbol filled out.
- Switchgear in the open position = Steady light in symbol empty.
- Circuit breaker tripped by a protective relay = Flashing light, symbol empty.
- Disconnector in motion = Flashing light, symbol empty or filled out according to direction of movement.
- Disagreement in position information = Steady light, symbol half filled out.

SAS diagrams also usually include all the information concerning the feeders at each voltage level, such as:

- Switchgear and auxiliary power system
- Related signals
- Change of status
- Measurements.

The values of the following variables are also shown:

- Voltages
- Currents
- Powers
- Temperatures.

In addition, SASs are provided with self-monitoring and diagnostic tools for critical parameters or functions. At the minimum, the following are monitored:

- Auxiliary supply voltages
- Availability of the various assemblies
- Circuits to transmission and execution of control commands
- Serial data communication links
- Software procedures
- Memories capabilities
- Timing periods
- Agreement of actual switchgear status and displayed status.

With respect to substation events occurring during operation (e.g., changes or updating of switchgear position, changes in alarm status, circuit breaker trips coming from protective relays), it is recommended that they are displayed on the screen and are also available in hard copy from a printer. Event messages must contain the following information at least:

- Event description
- Date and time of event
- Related voltage level
- Supplementary information.

The event list must record all events capable of producing significant consequences for the substation and related transmission lines in chronological order.

2.3 Alarming Function

In addition to the control of the substation from a local control room and main control house and the processing of the return confirmation signals from the switchgear, the SAS also processes and displays defect signals and other anomalous condition signals requiring acknowledgment, and forms the basis for deciding on what further action should be taken by the substation operator.

Typical alarm signals include:

- Operative unavailability of switchgear.
- Overloads or over-run of circuit breaker operating mechanism.
- Out of permissible range oil and/or winding temperature of the power transformer.
- Power transformer protective relay activated.
- Voltages values too high or too low.
- Loss of voltage.
- Failure of communication links.
- Failure of station/bay controllers.
- Failure of protective relays.

Such alarm signals are usually grouped in to major and minor categories according to their possible impact on the substation performance. To display alarm signals, in addition to the control system screen, an external device (alarm-annunciator) may be also installed. This component allows improved alarm perception and acknowledgment by the substation operator.

2.4 Measurement Function

Another significant function of SASs corresponds to measurement. This provides updated information needed for adequate operation of the power system. The main parameters subjected to measurement include:

- Active powers
- Reactive powers
- Voltages
- Currents
- Temperatures on power transformers.

The values are usually shown by measurement dialog windows dedicated to each voltage level of the substation.

2.5 Setting and Monitoring of Protective Relays

Protective relays dedicated to different applications (busbar protection, transmission lines protection, power transformer protection, etc.) work based on reference parameters that define when the relay is ready to deliver a selective trip command to a circuit breaker. The power system operator can use SAS facilities for fix or change those reference parameters.

2.6 Control and Monitoring of the Auxiliary Power System

Substations have internal electrical loads that require AC power service, like motors of circuit breakers and power transformers, lighting circuits and air conditioned equipment. This also requires DC power sources to provide electricity to IEDs and other secondary devices.

These internal power needs are covered by the auxiliary power systems conformed by a well configured low voltage network integrated of distribution transformers, transfer switches, LV cubicles, batteries and also sometimes diesel generators. The functions for control and monitoring of such systems nowadays are completely incorporated into SAS solutions.

2.7 Voltage Regulation

Electrical apparatus are manufactured to perform at a steady state at nominal voltage plus or minus a certain voltage percentage, for example 5%. When the applied voltage exceeds the limit value, damage may occur and this mainly affects the internal insulation. Also, when applied voltage is lower than the allowed value, malfunction can be experienced.

Voltage regulation is the process that maintains the power system voltage in the permitted range. It changes the winding turn number of power transformers through a transformer subsystem called a *tap changer*. Control commands and signals associated with such a subsystem and managed by the SAS include:

- Control commands to modify tap-changer position.
- Tap-changer position indication.
- Alarms due electrical or mechanicals faults on all the chain of voltage regulation process.

Further Reading

Arora, R. and Mosch, W. (2011) *High Voltage and Electrical Insulation Engineering*, Wiley-IEEE Press, Chichester.

Cadick, J., Capelli-Schellpfeffer, M. and Neitzel, D. (2005) *Electrical Safety Handbook*, McGraw-Hill, New York.

Degerfalt, M. and Herriman, G. (n.d.) The Homogeneous SA "Substation Automation" Solution, paper by ABB Inc. Canada.

Engmann, G., Fletcher, P., Luigi, G., et al. (2001) The substation design process – An overview, *CIGRE SC23 Colloquium, Puerto Ordaz*.

Matei, G. (2008) New standard solutions for data acquisition in Intelligent Protection Systems, *U.P.B. Sci. Bull.*, Series C, 70(3) (available at http://www.scientificbulletin.upb.ro/rev_docs_arhiva/full9377.pdf, accessed May 26, 2015).

McDonald, J.D. (2012) *Electric Power Substations Engineering*. 3rd Edition, CRC Press, Boca Raton.

Schlabbach, J. and Rofalski, K-H. (2014) *Power System Engineering: Planning, Design, and Operation of Power System and Equipment*, 2nd Edition, Wiley-VCH, Weinheim.

UCA International Users Group (2004) Implementation Guideline for Digital Interface to Instrument Transformers using IEC 61850–9-2 (Available at http://iec61850.ucaiug.org/implementation%20guidelines/digif_spec_9-2le_r2-1_040707-cb.pdf, accessed May 26, 2015).

Udren, E., Kunsman, S. and Dolezilek, D. (n.d.) Significant Substation Communication Standardization Developments, paper by Cutler-Hammer, ABB Automation, Inc. and Schweitzer Engineering Laboratories, Inc.

Van Riet, M.J.M., Baldinger, F.L., Van Buijtenen, W.M., et al. (June 2005) Alternative approach for a total integrated secondary installation in MV Substations covering all possible and required functions, *CIRED 18th International Conference on Electricity Distribution*.

3

Impact of the IEC 61850 Standard on SAS Projects

Some people wrongly think that IEC 61850 is another communication protocol. The truth is, this standard is a well-structured and consistent series of publications (see Table 3.1) that defines a set of system requirements, specifies file formats for configuration of individual devices (also for the overall system in a standardized way) and establishes a family of communication subprocesses to be applied to a substation.

The defined system requirements comprise the following reliability points:

- The substation shall keep its operative condition even if a single SAS component fails.
- Redundant schemes of the SAS shall be designed in such way that single component failure does not give rise to loss of both redundant elements.
- Redundant SAS components shall be served from separate power sources.
- No failure mode shall produce erratic trip signals or other spontaneous control actions on primary switchgear.
- Three reliability class severities based on standard IEC 60870–4 are recommended (R1, R2 or R3). The applicable class for a specific project shall be agreed between user and vendor.
- The SAS vendor shall declare the MTTF of SAS components and subsystems.
- The protection function shall operate in an independent manner.

The specified file formats are created by using an object-oriented data model called SCL (Substation Configuration description Language), which is based on information referred to primary equipment and its arrangements, to the selected communication network, the definitive selected IEDs, as well as to specific logical elements and applications. Such files allow the exchange of configuration information in a flexible way, even when associated engineering tools from different manufacturers are used.

Substation Automation Systems: Design and Implementation, First Edition. Evelio Padilla.

Table 3.1 Existing parts of the IEC 61850 Standard

Standard IEC 61850 "Communication networks and systems for power utility automation" set of publications			
Number	Title	Edition	Date of Publication
61850–1	Introduction and overview	2.0	2013–03
61850–2	Glossary	1.0	2003–08
61850–3	General requirements	2.0	2013–12
61850–4	System and project management	2.0	2011–04
61850–5	Communication requirements for functions and device models	2.0	2013–01
61850–6	Configuration description language for communication in electrical substations related to IEDs	2.0	2009–12
61850–7-1	Basic communication structure – principles and models	2.0	2011–07
61850–7-2	Basic information and communication structure – Abstract Communication Service Interface (ACSI)	2.0	2010–08
61850–7-3	Basic communication structure – Common data classes	2.0	2010–12
61850–7-4	Basic communication structure – Compatible logical node classes and data object classes	2.0	2010–03
61850–7-410	Basic communication structure- Hydroelectric power plants – Communication for monitoring and control	2.0	2012–10
61850–7-420	Basic communication structure – Distributed energy resources logical nodes	1.0	2009–05
61850–7-510	Basic communication structure – Hydroelectric power plants – Modeling concepts and guidelines	1.0	2012–03
61850–8-1	Specific Communication Service Mapping (SCSM) – Mapping to MMS (ISO 9506–1 and ISO 9506–2) and to ISO/IEC 8802–3	2.0	2011–06
61850–9-2	Specific Communication Service Mapping (SCSM) – Sampled values over ISO/IEC 8802–3	2.0	2011–09
61850–10	Conformance testing	2.0	2012–12
61850–80–1	Guideline to exchange information from CDC-based model using IEC 60870–5-101 or 60870–5-104	1.0	2008–12
61850–90–1	Communication between substations	1.0	2010–03
61850–90–4	Network engineering guidelines	1.0	2013–08
61850–90–5	Use of IEC 61850 to transmit Synchrophasor information according to IEEE C37.118	1.0	2012–05
61850–90–7	Object models for power converters in distributed energy resources (DER) systems	1.0	2013–02

The main established communication subprocesses, formally known as communication services, are the following.

GOOSE messaging (Generic Object Oriented Substation Event) This is a communication method that is able to distribute time critical information to several IEDs simultaneously, such as switchgear status indications, control commands and blocking signals. The method is based on the publisher/subscriber communication principle in which both IEDs, sender and receiver, use a local buffer to handle the process of data exchange.

MMS (Manufacturing Message Specification) This communication service is used to transmit medium priority information. As was defined by the Standard ISO 9506 some years ago, this communication method works under the client/server concept providing features for information exchange between nodes on a communication network with the advantages of interoperability among IEDs supplied by different vendors, independent of the type of application and with significant efficiency of obtaining and distributing the required information.

SV (Sampled Values) This communication service, considered in the standard for transmit data that comes from instrument transformers, uses the publisher/subscriber principle for communication in such a way that the publisher device (merging unit) records the acquired value in an internal buffer and the subscriber device (IED) reads the value and transfers it to its own buffer. A time stamp is added to each acquired value in order to allow the subscriber device to follow the right value sequence.

3.1 Impact on System Implementation Philosophy

The IEC 61850 standard gives substation owners the opportunity to re-engineer some of their traditional views to implement SAS functions (control, protection, monitoring, etc.). This bring not only economic advantages, but also the possibility to install cubicles, devices and communication media in a clean manner, improving secondary system reliability and providing a safer environment for maintenance works. Utilities and other users are already making use of that opportunity by defining audacious solutions, such as innovative protection schemes based on emerging IED features or relocating protection functions using benefits derived from increasing technological progress in the development of merging units used at system process level.

3.2 Impact on User Specification

Although a large amount of literature recently written is about features, benefits and practical applications of Standard IEC 61850, utilities and other users are still concerned about the best way to specify a SAS according to such a standard. This is very important in order to obtain a technically acceptable solution at the end of the project and also to avoid ambiguous scenarios during project execution that may affect project costs and/or timescales. On this matter, three major subjects shall be covered in the user specification.

The first aspect referred to is wanted functionality, because although most of the SAS solutions available in the marketplace are based on IEC 61850 and provided using current technology, IEDs are able to satisfy virtually all functions required for substation operation; some users may have reasons for preferring traditional schemes for specific functions. Considering that, the user specification must describe the functional scope of the SAS. This means that the user must indicate clearly all functions (control, monitoring, alarming, event managing, communication with remote ends, control and monitoring of auxiliary power system ...) that will be included into the system, as well as those functions that must be implemented through separate subsystems; for example, a dedicated subsystem for failure recording and processing. It is also important to make mention of any possible "exceptional" functionality that may be required by the user in line with their operational philosophy or standardized practices; for example, from which location is it possible to raise and lower positions of the tap changer of power transformers (main control house, local control room ...)? Or any particular provision, such as the recording and display of the number of circuit breaker operations.

The second valuable subject to be indicated in the specification is the set of conditions and values related to foreseen SAS performance (response times, safety provisions and reliability requirements). It may include the following:

- Redundancies in hardware and/or communication channels.
- Redundancy mode in HMI, if it applies (hot-hot, hot-standby).
- The number of bays or the feeder that shall be controlled by one IED.
- Certain operation principles, for example the select-before-execute principle applicable to control actions on switchgear.
- If any interlock override facility shall be provided in order to bypass the interlocking function under emergency conditions.
- Time limits for function execution, such as the maximum time allowed for completing a control command on primary switchgear.
- If control function and protection function are to be carried out by separate IEDs.
- Minimum capabilities for data storage in SAS components (e.g., 300 events or alarms).
- Updating the rate of analog signals coming from instrument transformers, for example every 1 s.
- The requested reliability class severity (R1, R2 or R3).
- Time resolution for time-tagging of events and alarms, for example 1 ms.
- Maximum load of processor and RAM under normal operation, for example 40%.

Last but by no means least, an important topic that must be indicated in the specification is the group of constraints imposed by the user, which may include the following:

- If the auxiliary contacts of the circuit breakers and disconnectors are wired up to IEDs located in the control room, in place this allows the use of actuators and sensors as considered by the standard.
- If the low voltage contacts of instrument transformers are connected directly to IEDs without the need for interposing the merging units considered by the standard.
- If any specific communication is made through serial communication, for example the communication between control and protection devices.
- If the interlocking logic is implemented by hard-wired media, instead this allows the vendor to do that through a GOOSE message defined in the standard.

3.3 Impact on the Overall Procurement Process

Most substation owners have their standardized procurement mechanisms based on accumulated experience from past projects and supported by their classical organizational structures. The introduction of IEC 61850 means a challenge for utilities and other SAS users, because the new technology associated with that standard impacts strongly on traditional corporative culture and practices, such as the form to implement technical solutions, the professional profile of employees, the motivation for continuous learning, the ability to conform teamwork and the validity of some work methods.

The main impacts identified in this respect comprise the following:

- SAS user has to induce and support employee education referred to IEC 61850, in order to gain strength in having effective actors in the procurement process and the rest of the project stages.
- A special multidisciplinary teamwork may be needed to discuss benefits and to allow implementation of full IEC 61850 functionalities, such as interlocking logic by a GOOSE message or the use of merging units for getting information from instrument transformers.
- The standard provides the possibility of increasing the amount of actors on SAS projects, for example several IED vendors. This means that the figure of a well-qualified system integrator becomes a key factor in the project success.

Also, beyond the impact of the IEC 61850 itself, changes in qualification and attitudes of user personnel are also useful when considering the emerging technological trends and nature of digital systems. This includes the multi-skill professional profile expected from engineers and technicians involved with a SAS project and also improved time responses for making decisions and for providing information.

3.4 Impact on the Engineering Process

SAS engineering process covers the following activities:

- Design and definition of a satisfactory solution.
- Configuration of selected devices and the overall system.
- Parameterization of selected devices.
- Preparation of engineering drawing and other technical documents.

The IEC 61850 Standard affects all of these activities, starting from the design stage, wherein defined reliability requirements must be followed; devices, system configuration and parameterization are now made through SCL files, and engineering drawing becomes simpler, mainly showing communication ports of IEDs instead of the old fashioned engineering drawing based on lots of connections between physical contacts. This means that the standard gives the opportunity to optimize the entire engineering process.

3.5 Impact on Project Execution

Apart from the cost saving associated with past generations of digital SASs due to a reduction in physical copper wiring, the adoption of Standard IEC 61850 allows an increase in such an economic impact by reducing even more metallic cabling if interlocking logic is implemented by using GOOSE messages and the analog values coming from instrument transformers are incorporated into the system by using merging units recommended by the standard. Additional cost savings are also derived from reduced effort and time required during site acceptance testing stage of the project, due the benefits of using the already mentioned SCL files, which also reduces the possibility of human error.

3.6 Impact on Utility Global Strategies

The standard allows an increasing and irreversible penetration of digital systems in the substation and beyond. This forces utilities to face up to several corporate factors that include the following:

- Substation becomes strongly based on fast changing technologies. This reality may be incompatible with the traditional trend of utilities in terms of systems standardization.
- The integration tendency in function/devices calls for a revision of professional staff profiles, in particular the old figures of dedicated protection engineer or control engineer.
- New SAS functionality includes information and means useful for replacing time-based maintenance with the more reasonable condition-based maintenance practice (e.g., number of operations of primary switchgear and self-supervision of IEDs).

In this respect, utility management is encouraged to implement technical and corporative strategies in order to improve efficiency in dealing with SAS projects and to gain the full spectrum of benefits offered by the standard.

3.7 The Contents of the Standard

As was shown before in Table 3.1, the Standard IEC 61850 is comprised of a set of separate publications in which titles may not be clear enough with regard to their contents. This may cause doubt in prospective readers when they face the task of learning about the Standard. To avoid that, a summary of the contents is indicated in Table 3.2.

3.8 Dealing with the Standard

Study and understanding of Standard IEC 61850 is a "hot" issue nowadays, but some people may be disappointed when they try to read part of it, particularly those professionals who are not deeply involved with IT/communication technologies, such as power engineers and other professionals who have graduated in classic electrical branches of study. This is due to the strong content of abstract models and specialized functional principles in the standard.

Furthermore, the amount of existing standard parts may also be intimidating to none expert readers. However, depending of the role of the reader in specific SAS projects, there are a small number of standard parts, as suggested in Table 3.3, which must be studied as a priority.

Table 3.2 Content of the Standard

Standard IEC 61850 "Communication networks and systems for power utility automation" Content of different parts		
Number	Title	Content
61850–1	Introduction and overview	Define the communication between IEDs. Define application scope of the standard.
61850–2	Glossary	Contains specific terms used in different parts of the standard.
61850–3	General requirements	Establish SAS quality attributes like: reliability, availability and security. Define applicable environmental conditions.
61850–4	System and project management	Describe various topics involved with SAS projects such as: engineering requirements, life-cycle concept and quality assurance practices.
61850–5	Communication requirements for functions and device models	Establish a standard SAS structure. Define station bus and process bus. Declare interoperability as the goal of the standard. Introduce the logical node concept and applications.
61850–6	Configuration description language for communication in electrical substations related to IEDs	Specify general standard file format (SCL) for communication purpose. Define various configuration Tools. Establish substation, IED and communication models. Define specific types of SCL files.
61850–7-1	Basic communication structure – principles and models	Covers modeling concepts and descriptions for device functions, historical data, control hierarchy and time synchronization.
61850–7-2	Basic information and communication structure – Abstract Communication Service Interface (ACSI)	Provides communication models for information exchange between IEDs.
61850–7-3	Basic communication structure – Common data classes	Specify data management principles related to status information, measuring information and setting facilities.
61850–7-4	Basic communication structure – Compatible logical node classes and data object classes	Specify device modeling and names of logical nodes for communication between IEDS.
61850–7-410	Basic communication structure- Hydroelectric power plants – Communication for monitoring and control	Specify data model and logical nodes for use in control facilities of hydropower plants.

(*continued*)

Table 3.2 (*continued*)

Standard IEC 61850 "Communication networks and systems for power utility automation" Content of different parts		
Number	Title	Content
61850–7-420	Basic communication structure – Distributed energy resource logical nodes	Define information for modeling principles applicable to dispersed generation facilities including micro-turbines and photovoltaic installations.
61850–7-510	Basic communication structure – Hydroelectric power plants – Modeling concepts and guidelines	Explain how to use logical node concept in control functions for power plants.
61850–8-1	Specific Communication Service Mapping (SCSM) – Mapping to MMS (ISO 9506–1 and ISO 9506–2) and to ISO/IEC 8802–3	Specify methods of exchanging data into Station Bus by using MMS and GOOSE communication services.
61850–9-2	Specific Communication Service Mapping (SCSM) – Sampled values over ISO/IEC 8802–3	Specify methods of exchanging data through process bus by using the sampled values technique.
61850–10	Conformance testing	Specify testing techniques for conformance checking and measurements techniques to declare performance parameters.
61850–80–1	Guideline to exchange information from CDC-based model using IEC 60870–5-101 or 60870–5-104	Provide guidelines to exchange information between substations and control centers.
61850–90–1	Communication between substations	Provide guideline on communication services and communication architectures.
61850–90–4	Network engineering guidelines	This part is addressed mainly to device vendors and system integrators. Outline engineering considerations including advantages and disadvantages of various network topologies and redundant schemes.
61850–90–5	Use of IEC 61850 to transmit Synchrophasor information according to IEEE C37.118	Provides techniques for transferring Synchrophasor data from Phasor Merging Units (PMUs) to Wide Area control and monitoring networks.
61850–90–7	Object models for power converters in distributed energy resources (DER) systems	Describe functionalities for systems associated to voltage converters.

Table 3.3 Suggested priority reading in the Standard

Standard IEC 61850 "Communication networks and systems for power utility automation" Priority in reading by different SAS project actors						
Part Number	Role in the SAS project/activity					
	System Specification	System Design	Project Lead	System Integrator	Site Testing	Maintenance
61850–1	x	x	x	x	x	x
61850–2	x	x	x	x		
61850–3	x	x	x	x	x	x
61850–4	x	x	x	x		
61850–5	x	x	x	x	x	x
61850–6		x		x	x	x
61850–7-1		x		x		
61850–7-2		x		x		
61850–7-3		x		x		
61850–7-4		x		x		
61850–7-410						
61850–7-420						
61850–7-510						
61850–8-1		x		x	x	x
61850–9-2		x		x	x	x
61850–10	x	x	x	x		
61850–80–1	x	x		x		
61850–90–1						
61850–90–4		x		x		
61850–90–5						
61850–90–7						

Further Reading

Baigent, D., Adamiak, M. and Mackiewicz, R. (n.d.) *IEC 61850 Communication Networks and Systems in Substations: An Overview for Users*, paper by GE Multilin and SISCO Inc.

Borroy Vicente, S., Gimenez de Urtasun, L., Villen Martinez, M., et al. (n.d.) *New protection scheme based on IEC 61850*, paper by C.P.S., Universidad de Zaragoza.

Botza, Y., Shaw, M., Allen, P., et al. (2008) *Configuration and performance of IEC 61850 for First-Time Users – UNC Charlotte Senior Design Project*, paper by University of North Carolina at Charlotte and Schweitzer Engineering Laboratories, Inc.

Brand, K-P., Brunner, C. and De Mesmaeker, I. (October 2005) How to use IEC 61850 in protection and automation, *Electra Review* 222, 11–21.

Brantley, R., Donahoe P.E.K., Theron, J., and Udren, E. (n.d.) *The Application of IEC 61850 to Replace Auxiliary Devices Including Lockout Relays*, paper by The Michigan Electric Transmission Company (METC).

De Mesmaeker, I., Rietmann, P., Brand, K-P. and Reinhardt, P. (2005) Practical considerations in applying IEC 61850 for protection and substation automation systems, *GCC Power Conference and Exhibition*, Doha, Qatar.

Dolezilek, D. (2005) *IEC 61850: What You Need to Know about Functionality and Practical Implementation*, paper by Schweitzer Engineering Laboratories, Inc.

Dolezilek, D., Whitehead, D. and Skendzic, V. (2010) *Integration of IEC 61850 and Sampled Value Services to Reduce Substation Wiring*, paper by Schweitzer Engineering Laboratories, Inc.

Engler, F., Kern, T.L., Andersson, L., et al., (2004) *IEC 61850 Based Digital Communication as Interface to the Primary Equipment*, paper B3–205, CIGRE session, Paris.

Hohlbaum, F., Hossenlopp, L. and Wong, G. (2004) *Concept and First Implementation of IEC 61850*, paper B5–110, CIGRE Session, Paris.

ISO 9506–1, Industrial Automation Systems – Manufacturing Message Specification – Part 1: Service definition.

ISO 9506–2, Industrial Automation Systems – Manufacturing Message Specification – Part 2: Protocol specification.

Kriger, C., Behardien, S. and Retonda-Modiya, J. (2013) A detailed analysis of the GOOSE message structure in an IEC 61850 Standard-based substation automation system, *Int J Comput Commun*, ISSN 1841–9836, 8(5), 708–721, South Africa.

Mackiewicz, R. (2005) *Technical Overview and Benefits of the IEC 61850 Standard for Substation Automation*, paper by SISCO, Inc.

Ozansoy, C. (2010) *Modelling and Object Oriented Implementation of IEC 61850*, Lambert Academic Publishing.

Patel, N. (2011) *IEC 61850 Horizontal GOOSE Communication and Overview*, Lambert Academic Publishing,

Ren, Y., Xiao, Y., Jin, Y. and Peng, S. (2013) Impact of IEC 61850 on substation design, *Journal of International Council on Electrical Engineering* 3(3), 210–214, China.

SISCO Inc. (1995) Overview and Introduction to the Manufacturing Message Specification (MMS).

4

Switchyard Level, Equipment and Interfaces

Most large substations have four different physical environments: The main control house, one or more local control rooms, the switchyard and the voltage transformation area. The switchyard may have different appearances, mainly depending on the substation insulation media (air-insulated or gas-insulated) and its design for environmental exposition (outdoor or indoor). In case of air-insulated substations, the primary equipment are svelte structures made of porcelain or polymeric material that provide electrical insulation between the energized upper part and the substation earth plane. Some of this apparatus is filled with gas, such as puffer type circuit breakers, which contain sulfur hexafluoride (SF6). Others are filled with dielectric oil, such as instrument transformers and power transformers.

In the substation yard, switchgear, instrument transformers and other primary apparatus are mounted on metallic structures to achieve the minimum standardized safety distances for internal circulation and movement of people and vehicles, such as that shown in Figure 4.1. This chapter covers SAS related details of substation components installed in the switchyard.

4.1 Primary Equipment

In the lexicon of power systems, operating voltages applicable to equipment and devices have been classified as follows:

- *Low voltage*: When the three phase system voltage is less than 1 kV.
- *Medium voltage*: When the three phase system voltage is above 1 kV and less than 52 kV.
- *High voltage*: When the three phase system voltage is from 52 kV to less than 300 kV.
- *Extra high voltage*: When the three phase system voltage is from 300 kV to 765 kV.
- *Ultra high voltage*: When the three phase system voltage is above 765 kV.

Substation Automation Systems: Design and Implementation, First Edition. Evelio Padilla.

Figure 4.1 View of a 765 kV substation switchyard. Source: © Corpoelec. Reproduced with permission of Corpoelec

The primary equipment of a transmission substation operates at standardized rated voltages, such as 72.5, 123, 145, 245, 420, 525 and 765 kV, whilst in most distribution substations, primary equipment works at lower standardized voltages, such as 7.2, 12, 24 or 36 kV. Such equipment is manufactured in such a way that its insulation in service must reliably withstand predefined over-voltage levels; otherwise they can become severely damaged due to external flashovers or by internal insulation breakdown. Power system over-voltages include temporary over-voltages (also called industrial-frequency over-voltages) arising from anomalous events occurring in the power system such as earth faults, load rejections or ferro-resonance phenomena. SAS responsibilities include the protection of primary and secondary equipment from such temporary over-voltages.

With respect to thermal capability, primary equipment for transmission substations, which are connected in series in the power circuit (circuit breakers, disconnectors and current transformers), usually have a steady state rated current between 1200 and 3500 A. When any short circuit occurs in a substation or related transmission lines (accidental direct connection phase-to-phase or phase-to-earth), the circulating current across the equipment becomes extremely high, for example 30,000 A or more. The duration of such undesirable conditions is limited to a short time (e.g., less than 1 s) by protective relays belonging to the SAS in order to save equipment from burning or from any other destructive failure.

Furthermore, in addition to the high or medium voltage physical segment connected to the substation power circuit, primary equipment also has auxiliary switches or low voltage terminals needed to transmit signals that are required for the compliance of SAS functionality.

4.1.1 Switchgear

Although the word *switchgear* as used in the electrical industry can refer to a wide variety of arrangements conformed by medium/high voltage apparatus with its associates low voltage devices, in this chapter we use the term specifically to mean the primary equipment that acts as switches and/or isolators; for example, circuit breakers, disconnectors and earthing switches. A brief description of their constructive profiles and operating principles follows.

4.1.1.1 Circuit Breaker

Power Circuit Breakers (PCBs) are comprised of three identical poles mounted on a common frame or on individual supports. Each breaker pole consists of one or more sealed chambers where the current interrupting process takes place, plus an intermediate set of insulating rods and accessories to transmit the mechanical power needed to open or close the breaker power contact. The PCBs are equipped with powerful common or individual operating mechanisms able to permit successful performance under both circumstances: normal load and short-circuit conditions (see the central object in Figure 4.2).

The sealed chambers mentioned previously are designed to interrupt a current by forcible and safe separation of a set of contacts in the current path. That action creates an arc, which

Figure 4.2 Vertical style 400 kV circuit breaker. Source: © Corpoelec. Reproduced with permission of Corpoelec

must be extinguished quickly, safely and completely, in order to eliminate the flow of current. In most circuit breakers, where the rated voltage is 52 kV or above, the interrupting chambers are filled with SF6 gas at a certain over-pressure. During the CB opening process (usually around 50 ms), the arc expands and is exposed to the gas. This cools and extinguishes it.

Because the dielectric strength of such gas depends on density, SF6 circuit breakers are equipped with density switches consisting of temperature-compensated pressure switches that are able to close some auxiliary contacts for sounding alarms or blocking signals.

Depending on several factors, such as rated voltage, required breaking capacity as well as technological trend of the manufacturer, circuit breakers can be equipped with any of the following operating mechanisms:

- Pneumatic operating mechanism
- Hydraulic operating mechanism
- Spring operating mechanism.

All these operating mechanisms must be carefully maintained and monitored to warrant the correct performance of the CB. CBs are provided with one or several control cubicles, which usually contain a lot of devices and accessories such as the following:

- Trip coil 1 (first tripping circuit) (per pole, in case of separate operating mechanisms).
- Trip coil 2 (second tripping circuit) (per pole, in case of separate operating mechanisms).
- Closing coil (per pole, in case of separate operating mechanisms).
- Auxiliary switches reflecting of CB position (open or closed).
- Gas density monitor for SF6 gas supervision.
- Pressure switch for compressed air supervision (if applicable).
- Selector switch to choose local/remote CB operation mode.
- Push buttons for close/trip local commands.
- Compressor motor (if applicable).
- Spring set, hydraulic set or pneumatic set.
- Motor protection switch.
- Miniature circuit breakers.
- Heaters.

Some of these devices require AC sources from outside and this is the case for motors and heaters. The other devices, belonging to the control/monitoring branch, typically work with DC voltages.

4.1.1.2 Disconnector

In the open position, disconnectors (also called isolators or air switches) provide positive, visible air-gap isolation from other substation equipment and transmission line segments for safe examination, maintenance and repair. In the closed position, a disconnector must provide an adequate capacity to handle all normal and abnormal currents that flow in the associated power circuit.

Careful attention must be given to the principle that a disconnector cannot be operated unless the associated circuit breaker is in open position.

Figure 4.3 400 kV Disconnector in an open position. Source: © Corpoelec. Reproduced with permission of Corpoelec

As shown in Figure 4.3, disconnectors are composed of three identical poles mounted on common or separate frames, depending mainly on their rated voltage. Each pole consists of a set of movable conductive blades supported by a post insulator stack.

These apparatus are built in a variety of forms to accommodate the various requirements of electrical clearances and space limitations. For example, center break, double break, vertical break and pantograph.

Disconnectors are equipped with several devices including:

- Motor and gear drives (per pole, in case of separate operating mechanisms).
- Facilities for manual operation.
- Auxiliary switches reflecting of disconnector position (open or closed).
- Selector switch to choose local/remote operation mode.
- Push buttons for closing/opening local commands.
- Key arrangements for interlocking purpose.

4.1.1.3 Earthing Switch

These apparatus are designed for occasionally connecting to earth specific primary power circuits. They are used for earthing sections of a substation or associated transmission lines during inspections, maintenance work or repairs.

Frequently, they are attached on one or both sides of a disconnector, although they can also be built with their own separate base and insulator stack.

It is common that an earthing switch is mechanically interlocked with the disconnector on which it is mounted, to prevent both apparatus from being closed at the same time. Their accessories comprise a set of gear drives (generally manual) plus a lot of auxiliary switches for position indication and electrical interlocking purposes.

4.1.2 Instrument Transformers

Instrument transformers (ITs) are designed to supply electrical inputs to measuring instruments, meters, relays and other similar devices.

The basic types of IT are:

- Voltage transformers, which convert the primary voltage of a power circuit to standardized low and nonhazardous voltage levels.
- Current transformers, which convert current flowing through the primary power circuit to a proportional lower current value.

These apparatus allow separation of measuring devices and other secondary components from the high voltage side of the substation, providing a safe environment to deal with the entire secondary system.

4.1.2.1 Voltage Transformer

Voltage transformers can be constructed on the capacitive-divider principle or the inductive principle. The first, most popular method comprises a capacitor column inserted in porcelain shells filled with dielectric oil plus an inductive transformer unit installed at the foot of the apparatus (see Figure 4.4).

The capacitor's column serves simultaneously as a capacitive potential divider and a coupling capacitor. This means that a fraction of the capacitor stacks are connected in parallel with the primary winding of the transformer unit. One or more secondary windings of the transformer unit provide equivalent power circuit voltage (generally at 120 V) to all SAS components requiring that signal.

4.1.2.2 Current Transformer

Current transformers (CTs) are used to convert high voltage line current to a lower manageable standard value that is insulated from the substation power circuit. They are based on the inductive transformation principle. The primary winding consists of a conductive bar that is fed through the CT head. One side of the primary winding is normally insulated from the head. Around the primary bar one or more ring type magnetic cores are placed in which secondary windings are distributed to minimize secondary flux losses. CTs usually look like those shown in Figure 4.5.

Figure 4.4 Capacitive voltage transformers stored as spare parts. Source: © Corpoelec. Reproduced with permission of Corpoelec

Secondary windings are internally led out through oil-paper bushings to the CT base. All of them end in a secondary terminal box, such as that illustrated in Figure 4.6. From that box, all secondary devices that need current inputs are connected in a series secondary circuit. *Special care must be taken regarding short circuits and to connect to earth those CT secondary windings that are not in use.*

4.1.3 Power Transformers

Power transformers, the larger apparatus in a substation, are composed of a primary winding, a secondary winding and in some cases, a tertiary winding. These windings are constructed of copper or aluminum wires. All windings are centered on a common magnetic core made of steel sheets. The complete set of windings and magnetic core is immersed in a metallic tank filled with

Figure 4.5 Current transformers stored as spare parts. Source: © Corpoelec. Reproduced with permission of Corpoelec

Figure 4.6 View of a CT secondary terminal box. Source: © Corpoelec. Reproduced with permission of Corpoelec

dielectric oil to which a lot of accessories are attached and these are needed for optimal performance of the transformer. Construction options include the three phases inside a common tank and also separate single-phase units. All ends of windings are generally led out through condenser type bushings. For high voltage levels the finished apparatus looks like that shown in Figure 4.7.

Figure 4.7 View of a single-phase 765 kV power transformer. Source: © Corpoelec. Reproduced with permission of Corpoelec

Large power transformers are provided with cooling systems based on air/oil heat exchangers by natural or forced oil circulation means. They are usually fitted with on-load tap-changers as a facility to regulate secondary voltage.

Typical power transformer accessories include:

- *Pressure relief valve*: Essentially a spring loaded device designed to quickly reduce the pressure inside the tank to a normal value. Generally, a trip switch assembly is provided for signaling and monitoring purposes.
- *Resistance bulb*: This is a kind of bimetallic probe based on the variation of resistance as a function of temperature. When it is inserted in an electrical circuit with a measuring instrument, it can indicate the temperature of the transformer oil or winding.
- *Oil thermometer*: This element is used to for measuring the temperature of the dielectric oil and shows it outside via temperature indicators, as illustrated in Figure 4.8.
- *Winding temperature sensor*: This is a combined arrangement for obtaining winding temperature by an indirect method. It is based on the fact that the temperature difference between the winding and the oil in the transformer tank depends upon the current present in the winding. The device receives an input from an oil thermometer and adds the temperature effect produced by the winding current, which is obtained from a current transformer installed in the bushing of the power transformer.

Figure 4.8 Power transformer temperature indicators. Source: © Corpoelec. Reproduced with permission of Corpoelec

- *Buchholz relay*: Any internal trouble, such as insulation breakdown or high activity partial discharges, can cause decomposition of materials contained in power transformers, such as pressboard, wood and oil. This produces gas bubbles that rise to the top of transformer tank. The Buchholz relay is designed to detect such gas bubbles and activate switches for alarm or trip signals.
- *Oil level indicator*: The function of this accessory is to detect any abnormal up/down movement of the transformer oil level that may be caused by any oil leakage or internal faults. An alarm contact is provided to signal low oil level conditions.

4.1.4 Other Primary Equipment

As well as the already mentioned primary equipment, there are other important apparatus that form part of the substation primary plant, for example:

- *Surge arresters*: These apparatus consist of a single-pole stack of metal-oxide discs, which are used to protect the substation primary equipment by limiting transient over-voltages (such as direct or induced lightning over-voltages) and switching surge over-voltages produced by switchgear operations. After the impulse discharge they are able to interrupt the current resulting from the power circuit.

- *Shunt reactors*: A coil structure connected to the primary circuit for control of reactive power on extra/ultrahigh voltage power systems by compensating for the capacitive effect of long transmission lines. Constructive structure and associated accessories of shunt reactors are similar to those for power transformers.
- *Series reactor*: This less common apparatus, also of a coil form, is inserted between segments of the power system to limit short circuit current levels.
- *Shunt capacitor banks*: This installation consists of a lot of capacitor units mounted on racks in order to control reactive power on the power system.
- *Post insulator*: The simplest primary apparatus consisting of a single-pole isolator stack is used to support the busbars and flexible conductors needed to interconnect the rest of the primary equipment.

4.2 Medium and Low Voltage Components

In the substation yard some medium or low voltage elements are also installed, which are required for different functions. One of them is the outdoor auxiliary power transformer (see Figure 4.9) in which a primary winding is connected to an external distribution line or to a tertiary winding of a power transformer, and its secondary winding supplies power to the substation auxiliary system.

Figure 4.9 Medium voltage auxiliary power transformer. Source: © Corpoelec. Reproduced with permission of Corpoelec

Figure 4.10 A bay junction box. Source: © Corpoelec. Reproduced with permission of Corpoelec

Other components belonging to the substation secondary system are the bay junction boxes (see Figure 4.10), which in a traditionally constructed substation gather the extensions of all secondary windings from voltage transformers and current transformers installed in a bay.

Here, there are also configured voltage and current circuits for measuring instruments, relays and other secondary devices.

4.3 Electrical Connections between Primary Equipment

Primary switchgear and current transformers are generally integrated in repetitive modules connected to one or more three-phase sets of strips made of rigid or flexible conductors called *busbars*, which act as common electrical nodes in the substation. There are a number of standardized primary arrangements chosen by substation owners that take into account diverse aspects, such as operational flexibility and cost of the overall installation as well as particular reliability and availability requirements. Two of those typical arrangements are shown in Figure 4.11: a one-and-a-half breaker scheme is frequently used for 230 kV and above, and a double busbar single breaker scheme is used for 115 kV and lower rated voltages.

In a one-and-a-half breaker scheme under normal operational conditions, all circuit breakers are closed and both busbars are energized. When any incoming/outgoing circuit

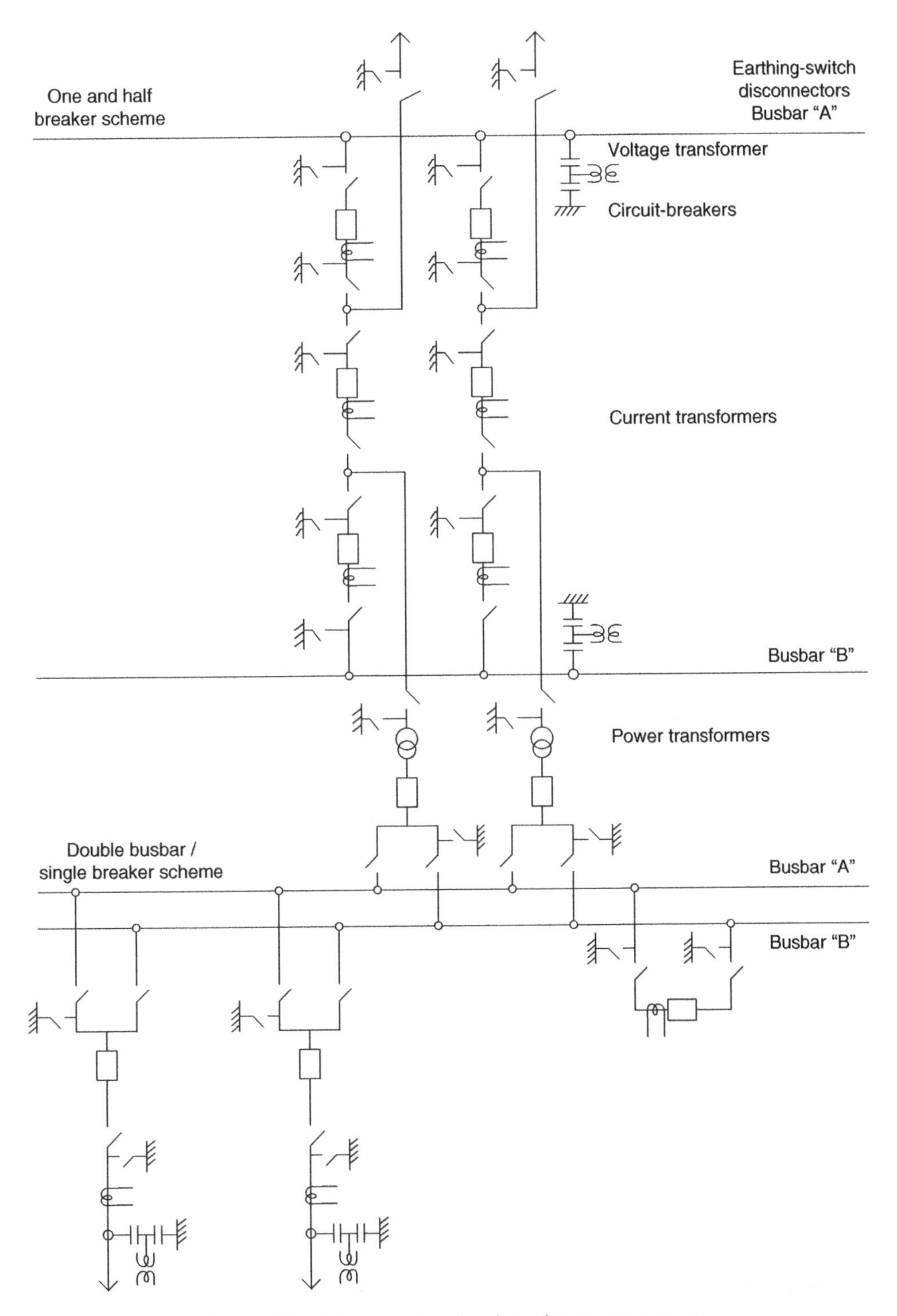

Figure 4.11 Example of a substation primary arrangement

faults, it causes the trip of the two associated circuit breakers and no other circuit is affected. This arrangement offers the following features:

- High availability of power supply.
- If a failure occurs on any busbar, it does not result in feeder disconnection from service.
- If a failure occurs on any circuit breaker installed on busbar sides, it does not result in feeder disconnection from service.
- Any busbar can be taken out of service at any time for maintenance work.
- Flexible for power system operative purposes and for maintenance work.
- The circuit breakers and current transformers must be able to handle the sum of the load currents of the two associate incoming/outgoing circuits.
- The implementation of relaying and automatic reclosing need complex configurations because the middle of the circuit breaker is responsible for functions on both associated circuits.

In a double busbar single breaker scheme, normally each circuit is connected to both the buses. When a feeder circuit breaker needs to be taken out for maintenance, it can be bypassed and the respective circuit protected by the coupler circuit breaker. This arrangement offers the following features:

- Each incoming/outgoing circuit may be connected to either busbar.
- A feeder's circuit breaker failure will take all the circuits connected to the busbar section out of service, which eventually means the entire switchyard.
- Either busbar may be disconnected for maintenance work.
- Allows an intermediate degree of flexibility.
- Circuit breaker maintenance may be carried out keeping all circuits in service.

4.3.1 Incoming Circuits

Incoming circuits, sometimes called line entrances, act as the power source to the substation. They may come from any generation station or from another remote substation. A view of an example is shown in Figure 4.12.

4.3.2 Outgoing Circuits

Outgoing circuits, also called feeders, leave from the power supply side transferring power to a distribution system or to any stream that is down-concentrated electrical load.

Typical physical elements served by outgoing circuits include:

- Line segments
- Distribution transformers
- Shunt capacitor banks
- Medium voltage cubicles.

Although most of those circuits serve as a radial link, care must be taken if there is the possibility of loop connection with other active circuits at a remote location.

Figure 4.12 A 765 kV line entrance. Source: © Corpoelec. Reproduced with permission of Corpoelec

4.3.3 The "Bay" Concept

In an SAS environment, a bay is understood to be a generally repetitive group of equipment within one voltage level, such as incoming/outgoing circuits belonging to a double busbar single breaker scheme or each one of the three breaker sets that make up a one-and-a-half breaker scheme (see Figure 4.11). The latter looks like the one shown in Figure 4.13.

4.4 Substation Physical Layout

The physical layout of a particular substation is strongly influenced by the selected primary arrangement in such way that, for all standardized primary arrangements, a corresponding classical physical layout exists. The usually implemented physical layouts for a one-and-a-half circuit breaker scheme and double busbar single breaker scheme are shown in Figure 4.14. The first is based on a three-level layout of primary conductors, while in the second, just two levels are enough.

In cases of apparent design trade-off situations, some variations on a typical layout have to be evaluated taking into account such factors as:

- Size and geometrical configuration of the land property.
- Maintenance requirements.
- Electrical clearances.

Figure 4.13 A 765 kV bay. Source: © Corpoelec. Reproduced with permission of Corpoelec

- Possibilities for control room locations.
- Needs for roadways.
- Aesthetic appearance of the substation.
- Type of primary conductor (rigid, flexible).
- Type of supporting structures.
- Environmental aspects.

In any case, the SAS designer shall know the definitive substation layout in advance (number of buildings, component locations, distances) in order to consider it when designing a SAS solution and device features.

4.5 Control Requirements at Switchyard Level

There is a number of control requirements at switchyard level needed to accomplish different activities during the life of the substation, such as indicated in Table 4.1.

The local operation of any of the switchgear must be submitted to the key interlocking system comprised of a group or series of interlocking devices applied to associated equipment in such a manner as to prevent, or allow, operation of the equipment only in a prearranged sequence. For example, to close an earthing switch, a key delivered by the associated line disconnector in open position is necessary. A sample of those interlocking devices is shown in Figure 4.15.

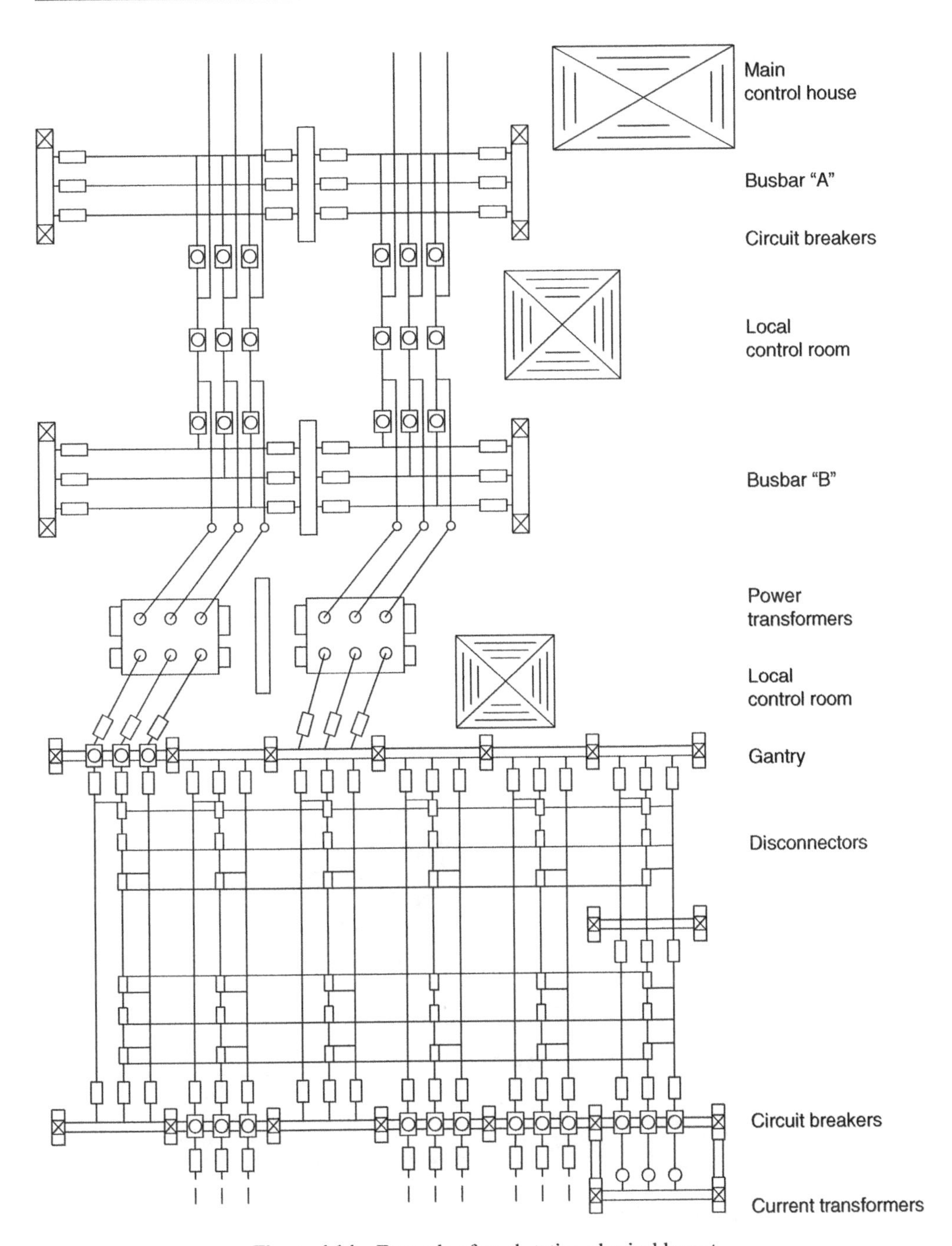

Figure 4.14 Example of a substation physical layout

Table 4.1 Switchyard control requirements

Apparatus	Control requirement/scenario	Control action
Circuit breakers	• Commissioning • Maintenance • Repairing	Local open/close command
Disconnectors	• Commissioning • Maintenance • Alignment of conductive blades • Repairing	Local open/close command
Earthing switches	• Commissioning • Maintenance • Repairing	Local open/close command
Power Transformers	Tap changer maintenance (if applicable)	Local up and down tap position

Figure 4.15 Set of key interlocking devices. Source: © Corpoelec. Reproduced with permission of Corpoelec

Further Reading

ANSI C 37.57 (n.d.) *Switchgear – Metal-Enclosed Interrupter Switchgear Assemblies* – Conformance Testing.

Bharat Heavy Electricals Limited (2007) *Handbook of Switchgears*. McGraw-Hill, New York.

CIGRE Working Group 23.03 (2000) General guidelines for the design of outdoor AC substations, Brochure 161.

Clerici, A., Ardito A., Eitzmann, M., et al. (1998) Temporary overvoltage withstand characteristics of extra high voltage equipment, *Electra Review* 179, 39–45.

Del Vecchio, R.M., Poulin, B., Feghali, P.T., et al. (2010) *Transformer Design Principles*. CRC Press, Boca Raton, FL.

Esmeraldo, P.C.V., Gaunt, C.T., Mader, D.J., et al. (1999) On behalf of CIGRE Task Force 6 of Working Group 33.11, Flashovers of open circuit-breakers caused by lightning strokes, *Electra Review* 186, 115–121.

Flurscheim, C.H. (2001) *Power Circuit Breaker Theory and Design*, IEE Power Engineering Series 1, The Institution of Engineering and Technology, London.

Greenwood, A. (1991) *Electrical Transients in Power Systems*, 2nd Edition, John Wiley & Sons, Ltd, Chichester.

Harlow, J.H. (2012) *Electric Power Transformer Engineering*. CRC Press, Boca Raton, FL.

IEC 60044 Standard, Instrument Transformers.

IEC 60071 Standard, Insulation Co-ordination.

IEC 60076 Standard, Power Transformers.

IEC 62271 Standard, High-voltage Switchgear and Controlgear.

IEEE Standard 80, Guide for Safety in AC Substation Grounding.

Karlsson, D., Olovsson, H.E., Wallin, L. and Solver, C.-E. (1996) *Reliability and life cycle cost estimates of 400 kV Substation Layouts*, IEEE paper PE-482-PWRD-0–11–1996.

Kuffel, J., Kuffel, E. and Zaengl, W.S. (2000) *High Voltage Engineering Fundamentals*. Elsevier B.V., Amsterdam.

Laughton, M.A. and Warne, D.F. (2003) *Electrical Engineers Reference Book*. Newnes (Elsevier Science), Amsterdam.

O'Connell, P., Heil, F., Henriot, J., et al. (2001) SF6 in the Electric Industry – Status 2000, *CIGRE SC23 Colloquium*, Puerto Ordaz.

Rizk, F.A.M. and Trinh, G.N. (2014) *High Voltage Engineering*. CRC Press, Boca Raton, FL.

Schavemaker, P. and van Der Sluis, L. (2008) *Electrical Power System Essentials*. John Wiley & Sons, Ltd, Chichester.

Smeets, R., van der Sluis, L., Kapetanovic, M., et al. (2014) *Switching in Electrical Transmission and Distribution Systems*. John Wiley & Sons, Ltd, Chichester.

Wahlstrom, B.H.E. (1999) On behalf of CIGRE Working Group 33.10, Temporary overvoltages – System aspects, *Electra Review* 185, 89–99.

5

Bay Level: Components and Incident Factors

The bay level, as defined by IEC 61850 standards, is an intermediate control place between switchgear boards and the main control house of the substation. At this level, the control, monitoring and other important functions are performed from one of several local control rooms located into the switchyard. Those control rooms, usually at least one per each primary voltage of the substation, lodge the majority of IEDs, relays and other secondary devices belonging to the SAS. Devices are mounted on racks inside metallic cubicles generally individualized by function: control, protection, measurement and so on. From here, IEDs connect with primary equipment installed at the switchyard through copper wiring or by an emerging network arrangement called the *process bus*. This chapter deals with bay level components and the means to ensure survival in the surrounding environment.

5.1 Environmental and Operational Factors

Substation equipment and devices are exposed to aggressive natural phenomena such as lightning strikes and seismic action, as well as to severe transient waves and other electromagnetic disturbances caused by various events such as the switching of reactive loads and phases to earth insulation faults. These facts bring about the necessity for substation engineers to understand the theory basis of certain phenomenon and the way in which they may affect substation components.

5.1.1 Lightning Strike

Lightning strike is essentially a gigantic spark (technically, a large electrostatic discharge) traveling from a cloud to the Earth's surface (mostly, lightning strikes from cloud to cloud) and dissipating enormous amounts of energy in just a few milliseconds. The phenomenon

---

Substation Automation Systems: Design and Implementation, First Edition. Evelio Padilla.

results in a peak value of discharged DC type current than can reach 50,000 A or even more, developing electrical potentials of 30 million V or greater.

Besides the risks to people and animals, lightning strike also may cause harm to physical structures by way of fire and burning. Lightning strike may affect a substation by way of direct impact or through impacts on associated transmission lines. When lightning strike occurs at the substation or nearby, the short duration but high intensity current tends to tear or bend metal parts, while a strike on insulating or semi-insulating elements can produce explosive reactions capable of causing significant damage.

Current flow through poor conductive capability paths, like a metallic route with high electrical impedance, induces local voltage gradients that may generate dangerous sparks (side flashing) and also may create potential differences between points to which a person may be in contact (touch voltage) big enough to cause shock injuries. Potential difference can also build up in the soil across the span of the step a human can take (step voltage), which may also be lethal.

When lightning impacts on any of the associated transmission lines, generally the current and the instantaneously developed voltage travel as waves from the point of contact in both directions along the shielding wire. Those waves face up surge impedance discontinuity rapidly at adjacent towers from where they reflect successively in such a way that a large number of additional waves are generated in the short term. During that reflecting process, the voltage created on the shielding wire may switch to the opposite voltage polarity of any of the phase conductors. This may produce a particular electric discharge called *back flashover*, which is characterized by a significant content of high frequency transients that are able to reach substation equipment.

Lightning strike on transmission lines, even if it impacts the shielding wire first, may also induce surge voltages on phase conductors by either the mechanism of electrostatic induction or electromagnetic induction. That induced voltage surge may, in some cases, arrive at a substation with a significant amplitude.

As a lower probability event, the lightning can also impact directly to one or several phase conductors (due to shielding failure). In such cases, a lightning impulse may reach the substation carrying the complete amount of dissipated energy.

5.1.2 Switching Transients

Unlike lightning waves, switching transients are self-induced by the operation of disconnectors and circuit breakers. The reason those transients exist is because of the intrinsic inductive and capacitive proprieties of power system components. They are composed of relatively slow shape impulse type voltages called switching surges and sometimes also of extremely high frequency oscillatory currents and voltages, best known as fast transients.

A greater concern than switching surge voltages, which are limited by surge arresters, is fast transients because of the high frequencies that can damage insulation internal structures of primary equipment and damage to or malfunctions in secondary devices.

The mechanisms of switching transient generation can be described approximately as follows:

5.1.2.1 Disconnector Operation

As already mentioned, this equipment is usually composed of three separate poles, one per phase, sometimes mechanically fixed to a common base. When it is undergoing the closing operation, which may take some seconds, a series of arc phenomena occur between the moving contacts. These arcs produce current transient waves on the busbar, which travel away from the disconnector in both directions. Due to the mechanical nature of the operation mechanism, not all three poles operate at the same time. This means that during the phase associated with the pole that operates first, the initial transient appears. Later, the rest of the poles start to act as secondary sources of transients. Studies show that secondary transients are more intensive when compared with the initial transient. The higher amplitude value of the transient current can be seen at the beginning of the closing process. This is due to the distance between contacts being greatest at that point and a strong voltage gradient is required to produce arcing.

When the disconnector is opening, the transient current shows the inverse behavior, for example the current amplitudes are small at the beginning and they increase as time passes.

In both cases, the electric arc ignites and extinguishes multiple times generating hundreds of transient currents characterized by their very high frequency harmonics, in most cases in the range of 50 kHz–1 MHz, which also produce significant high frequency electric and magnetic fields in the proximity of the operating disconnector.

5.1.2.2 Circuit Breaker Operation

In the case of circuit breaker operations, the term of transient generation becomes shorter because the circuit breaker operation process usually takes less than 100 ms. When the circuit breaker is in the closing operation, transients are produced due to pre-strike phenomena before the moving contacts reach physical contact. This is more severe in cases of re-energizing long transmission lines in UHV substations or the energizing of capacitor banks, especially if another capacitor bank is connected to the same busbar and damping reactors are not installed.

The opening of a circuit breaker also produces transients generated by electric discharge across small distances between moving contacts before the arc is extinguished. It is particularly important when re-ignition phenomena occur, such as in the case of applying a circuit breaker to switching inductive loads, such as shunt reactors in UHV substations.

5.1.3 Electromagnetic Disturbance Phenomenon

Transients present in the primary power circuit of the substation may be transferred to the circuits of the secondary system through inductive, capacitive and conductive coupling mechanisms. Inductive coupling is due to mutual inductance that appears between HV conductors and low voltage conductors located on the earth plane. Capacitive coupling comes from parasitic capacitances also between HV conductor and low voltage conductors.

Conductively coupled transients are caused mainly when high frequency current flows from the primary power circuit to the earth through stray capacitance in the HV apparatus, especially in capacitive voltage transformers. During this event, the potential increases locally at the point where the current injection takes place because of the inherent surge impedance, which the current faces, going up before arriving to the earth (effect well known as Transient Earth Potential Rise TEPR or TGPR).

All of these transferred transients suggest electromagnetic disturbances consisting of sudden variations in voltages or currents that are able to reach secondary devices via any of their physical ports (power supply, communication, input/output and earthing).

Other electromagnetic disturbances can be also propagated by radiation in the form of electromagnetic waves emitted from HV primary circuit when transients are present. These disturbances can affect secondary devices by destruction of their internal sub-assemblies, or by interfering with their operation by either loss of operation or false operation.

In recent years, there has been increasing concern surrounding this issue, due mainly to disturbances getting stronger with the increase in substation primary voltages and currents, while modern secondary devices are becoming increasingly sensitive.

5.1.4 Lightning Protection Practices

Like any other outdoor installation, substations are exposed to direct lightning strikes, in addition to lightning impulse waves coming from associated transmission lines through other paths, line overhead shielding wires or line phase conductors.

Lightning protection systems designed for direct impact are built to protect personnel and equipment/devices from hazards arising from lightning strikes. In a more specific manner, such a system must be designed for the following:

- Protection of people in outdoor environments from direct impact of lightning strikes.
- Protection of people in indoor environments where they may be at indirect risk due to lightning currents circulating into the buildings.
- Protection of buildings, structures and primary equipment.
- Protection of sensitive electronic devices from disturbances resulting from lightning strikes.

Those goals are traditionally achieved by providing the substation with an "electrostatic umbrella", such that a direct strike on the substation terminates on any overhead lightning protection conductor and a good conductive route allows lightning currents to flow safe to earth. This shielding system consists of vertical mounted lightning masts (also called lightning rods, Franklin ends or air terminals) and also, if necessary, a horizontal arrangement of overhead shielding wires. Lightning masts, which tend to be preferred by substation owners because of they provide a greater security of protection service and are easy to maintain, are usually installed on the top of gantries or other steel-column constructions.

Overhead shielding wire arrangements (Figure 5.1) are recommended to be configured as closed loops interconnected with lightning masts and connected to earth by down-low impedance conductors by way of at least two separate earthing connections.

Although the previously mentioned shielding system fulfills an essential function by protecting a substation against direct lightning strike, no less important are the means to also protect substation components from lightning waves coming from associated overhead transmission lines through phase conductors. The classical solution to that is by installing surge arresters as voltage limiting elements as close possible to more expensive and complex equipment like power transformers and shunt reactors (see Figure 5.2). Surge arresters are, in some cases, also installed at the line entrances, depending on results

Figure 5.1 Substation provided with overhead shielding wires. Source: © Corpoelec. Reproduced with permission of Corpoelec

Figure 5.2 Power transformers protected by surge arresters. Source: © Corpoelec. Reproduced with permission of Corpoelec

of power system studies considering several factors such as isokeraunic activity on the geographical location (thunderstorm days per year) and the expected affectivity of overhead line shielding wires.

Even if careful designs are made to define the characteristics of shielding systems and surge arrester selection, to get an effective overall lightning protection system, it is essential that down-conductors from both solutions terminate connected to a well-designed earthing system, the details of which are explained in the next section.

5.1.5 Typical Earthing Systems

An earthing system plays a vital role in all electrical systems. In substations, it acts as a large electrode for providing the zero potential reference and for draining fault currents to earth, while also forming part of the entire group of safety facilities. The main reasons for installing an earthing system at a substation include the following:

- To ensure safety to operating personnel by limiting dangerous touch and step voltages even at highest earth fault currents.
- To provide the earth connection for the neutral/earth terminals of power transformers, voltage transformers, shunt reactors and capacitor banks.
- To allow the right performance of surge arresters or other similar protective means.
- To discharge prior to (and keep safe primary equipment and secondary devices during) maintenance or repair activities.
- To provide a sufficiently low impedance path when high frequency currents flow, in such way that a minimum earth potential rise occurs.

The design process of an earthing system is generally carried out under a trade-off scenario to ensure that safety and operational issues are taken care of without constructing an over-dimensioned, more expensive than necessary system. The most relevant design parameters are: Fault current value, fault duration and soil characteristics.

Maximum fault current values given by power system studies consider single-phase and three-phase faults, and include possible future increases due to voltage up-rating of associated transmission lines as well as the effect of future new power generation plants in the surrounding region. As to fault duration, a back-up clearance time of protective relays is usually recommended. Soil characteristics come from resistivity tests normally performed by the substation owner.

Physically, the substation earthing system consists of an integrated mesh arrangement formed of earth electrodes, earth conductors and protective bonding conductors. Earth electrodes are vertically buried rods, generally made up of copper coated steel material resistant to corrosion effects throughout the entire lifetime of the substation. Earth conductors, typically made up of copper stranded conductors, form a reticulated earth mesh that is buried around 60 cm below the soil surface, such as shown in Figure 5.3. The optimized geometrical configuration of the mesh, the total length of the buried conductor and the minimum acceptable diameter of the earth conductor, results from extensive calculations performed for each particular substation project. Protective bonding conductors are also made up of copper stranded wires to connect equipment and structures with the earth mesh. Several of them are shown in Figure 5.4.

Figure 5.3 View of an already installed earth conductor. Source: © Corpoelec. Reproduced with permission of Corpoelec

Figure 5.4 Protective bonding conductors (right side). Source: © Corpoelec. Reproduced with permission of Corpoelec

5.1.6 Measures to Minimize Electromagnetic Effects

IEDs and other substation secondary devices may become victims of electromagnetic disturbances, mainly those caused by lightning strikes and switching operations. Undesirable effects on devices may be classified in the following four types:

1. Permanent and verifiable malfunctions.
2. Temporary malfunctions that occur only when a disturbance is present.
3. Rare malfunctions that occur sometimes when a disturbance is present.
4. Destructive failure due to internal damage.

Furthermore, due to the efforts of device manufactures with respect to how devices must be optimally built to minimize the probability of malfunctions/failures, and because the use of copper wiring remains a standard practice, a number of design-related and constructive mitigation methods implemented are still recommended today.

Earthing of Secondary Circuits Those secondary circuits that need a common voltage reference, for example the three circuits coming from secondary windings of a three-phase group of capacitive voltage transformers, must be jointed and earth connected to one single point in the earth system (usually inside bay junction box). This is to avoid a high frequency voltage between different points of the earth system inducing voltages in the circuits that reach secondary devices.

Constraining Secondary Circuits into Cables All wires belonging to a single secondary circuit must be contained within a single cable. In this way any loop is eliminated to avoid may create inductive coupling big enough to generate significant voltages in the circuit.

Functional Independence of Earth Mesh The earth mesh does not have to be used as the return path for any secondary circuit.

Relative Location HV Conductors – Secondary Cables Locations of secondary cables close to HV conductors have to be avoided. This is because the mutual inductance between HV conductor and cable is inversely proportional to the distance between them.

Cable Routing Different application cables must be routed separately. If a sufficient number of cableways are unavailable, power cables must be accommodated on one side of the cable duct and control cables on the other side, maintaining a minimum separation distance of 20 cm.

Screening of Control Cable This technique has proved to be effective at reducing the effects of electromagnetic disturbances, especially when the screen is well designed in terms of its screening factor and the thermal capability and also if the screen is connected to earth by reliable bonding leads at both ends of the cable.

Cubicle Earthing A typical and effective arrangement consists of bond to earth for all metallic cubicles through a common earth bus installed around the internal building walls and bonded to earth via a minimum of two conductors.

Building Earthing All building steelwork must be bonded to a common earth bus connected to earth at two points of the earth mesh at least.

Application of Fiber-Optic Cables Fiber-optic cables, immune to electromagnetic interference, are today widely used for communication between local control rooms and the main control house. It is also expected that downstream use of local control rooms will increase significantly in near future, mainly by using the process bus proposed by the IEC 61850 Standard.

5.2 Insulation Considerations in the Secondary System

Although insulation coordination, as a design aspect, can appear to only apply to the primary power circuit of the substation, the SAS is responsible for ensuring that the insulation of secondary equipment and devices is adequate to withstand the electrical stresses to which they are subjected.

Besides external incoming perturbations, such as those induced by lightning strikes and HV switchgear operations, there are also various internal events that produce significant over-voltages, for example MV/LV switchgear operations and faults on low and medium network. While an in-depth insulation coordination study is generally not justified, the substation design stage should include an assessment of impulse voltage levels that may appear on terminals of MV/LV equipment and devices. This estimation of over-voltage exposure applies on a case by case basis taking into account the following:

1. Origin of MV power supplies (tertiary winding of power transformer, overhead transmission lines …).
2. MV/LV switchgear types are going to be allowed (SF6, vacuum …).
3. The length and the type of connections between MV/LV elements.

Once the expected over-voltages are acquired, they are compared with insulation levels of equipment and devices to be installed, bearing in mind the convenience of including any kind of over-voltage protection. In addition, choosing the higher standardized value of basic insulation level (BIL) allowed by the standards, according the respective rated voltage of the equipment (particularly auxiliary power transformers), brings a relatively insignificant extra cost to gain additional safety margins in favor of substation availability and reliability.

5.3 Switchyard Control Rooms

These buildings (also called local control rooms) are designed to accommodate all the cubicles needed for the operation of the substation at the bay level. The cubicles must be arranged in a logical layout within the control room according to their function and contain the following:

1. Control and protection cubicles associated to the corresponding bays (main and back-up).
2. Teleprotection devices.
3. Time synchronization devices.
4. Power transformers, control and monitoring cubicles.
5. Capacitor bank, control and monitoring cubicle (if applicable).
6. Disturbance recorder cubicles.
7. Metering devices.
8. Communication equipment including interfaces.
9. Voltage regulating devices.
10. Fiber-optic distribution cubicle.
11. Auxiliary Services, control and monitoring cubicles.
12. DC auxiliary service cubicles.
13. AC distribution cubicles.

Figure 5.5 View of a switchyard/local control room. Source: © Corpoelec. Reproduced with permission of Corpoelec

Cubicle layout must be done in such way that operation staff can move freely inside the room.

As a constructive matter (see Figure 5.5), the substation owner usually specifies the minimum space requirement including provisions to lodge cubicles for future extension of the substation. This criterion allows the substation contractor to increase the definitive dimensions according to the chosen SAS solution and considers factors such as:

1. How many bays will be controlled by each single bay controller?
2. How many bay controllers will be allocated into a control cubicle?
3. How many protective relays will be installed into a single cubicle?
4. How much space will be required by other devices like Ethernet switches?

As a safety strategy, some utilities require two separate exits located at opposite sides of the control room to provide easily accessible means in case of an emergency.

With respect to additional means to protect secondary devices against electromagnetic disturbances, some utilities require installation of a shielding mesh (see Figure 5.6) on all walls of switchyard control rooms belonging to substations of 230 kV and above.

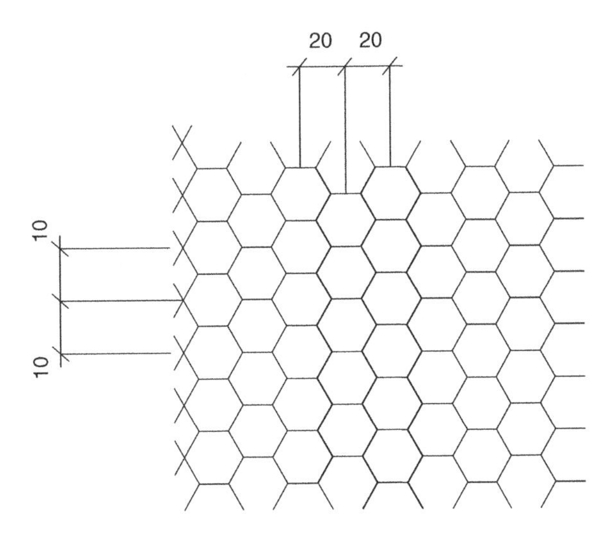

Figure 5.6 Mesh for shielding of switchyard control rooms

5.4 Attributes of Control Cubicles

The need for increased availability of HV substations requires high performance of both primary equipment and secondary components. Control cubicles (also called control panels, relay panels, cabinets or enclosures), as the container of IEDs, protective relays and other control-oriented devices, are required to fulfill exigent design and manufacturing criteria addressed at preserving device functionality, even under adverse or abnormal service conditions. Those proprieties are described in the following sections.

5.4.1 Constructive Features

This refers to the physical details ensuring that cubicles are able to properly accomplish their design objective, such as:

1. Structural frame strong enough to support devices and accessories.
2. Predominantly metal enclosed dust, moisture and vermin proof.
3. Degree of protection (according to the IEC Standard 60529) not less than IP-43.
4. Enclosure formed by sheet steel of thickness not less than 2 mm.

5.4.2 Earthquake Withstand Capability

In many geographical regions of the world earth tremors (seismic activity) occur due to accumulated stress in the soil and geological/tectonic plate activity. These earthquakes may cause damage to substation components, particularly those vertically configured apparatus with heavy loads on top.

Although earth acceleration caused by earthquakes reaches its highest value in the horizontal direction, vertical accelerations are also sometimes significant. The seismic severity at a particular site generally depends on geological data and seismic history.

The behavior of the control cubicle during an earthquake will depend on its own resonance frequencies, vibration modes and damping factors imposed by the structural integration, and mechanical loads associated with the internal components. Suitability of control cubicles to seismic requirements may be assessed by mechanical transfer studies and/or by tests. This suitability must be assessed early in the design stage, in order to avoid scenarios in which corrective measures result in more expense and are difficult to implement.

5.4.3 Electromagnetic Compatibility

The electromagnetic compatibility of a control cubicle represents its capability to keep its internal space free from the effects of external electric or magnetic fields while limiting its own field emissions to the surrounding environment. This is achieved by a high performance shielding enclosure usually characterized by the following factors:

- *Composition of the enclosure*: Use of conductive sheets outside and internal partitions.
- *Electrical continuity*: Providing continuous electrical contact between adjoining enclosure portions. Special care is needed to connect the door to the rest of the cubicle body through a low impedance tie with contact points free of any insulating coating.
- *Blocking the interference*: Protect all holes for cable routing, ventilation and accessories mounted on the front of the cubicle with special arrangements to stop EMI.

Further to design and constructive provisions, design tests should be done on cubicles to ensure the effectiveness of the shielding enclosure, according to current standardized procedures.

5.5 The Bay Controller (BC)

The bay controller, in practice an IED provided with a built-in type of human machine interface (also called the bay control unit or bay computer) consists of a hardware platform (electronic module) containing a software package (programming and data bases) configured to enable the implementation of various control functions, such as:

1. Control and monitoring of primary switchgear
2. Manage interlocking logics
3. Display alarm signals
4. Recording and display events information
5. Interface with protective relays
6. Synchrocheck for circuit breaker operation
7. Metering functions
8. Interface with upstream control facilities
9. Interface with other control devices like voltage regulators.

Physically, the bay controller is integrated with different elements chosen to suit particular requirements. Often they include the following:

- Housing with interconnection facilities.
- Power supply: Usually it is a DC/DC converter to avoid electromagnetic interference through this path. A second supply unit may be needed as supplementary or redundant.
- Communication module: Electro-optical converter and communication port to connect with the station bus.
- Processing unit: Electronic module in which application programs are installed provided of interfaces for local control, flash memory for data configuration and means to acquire information from primary equipment and exchange data with station controller.
- Interconnection module: Unit used to establish the internal communication between different elements of the bay controller.
- Communication controller: Module responsible for the external communication via station bus.
- Set of binary input cards: Generally conformed by optic-couplers to avoid electromagnetic interferences coming from secondary circuits.
- Binary output module: Often consist of auxiliary relays able to manage the power required for control command of primary and secondary switchgear.
- Instrument transformers modules: Units dedicated to receive voltage and currents signals coming from primary voltage transformers and current transformers.
- Signal converter module: Used for convert current and voltage analog signal into digital signals in such way that can be processed by application programs.

With respect to operating principles, the most common features include:

- Control cubicles (see Figure 5.7) are provided with a local/remote switch to allow authorized personnel execute control functions from the switchyard control room.
- Control commands from BC are made in a sequential way; the first step to select the particular switchgear and a second step for delivering the opening or closing control command.
- Alarm messages appear spontaneously.
- Event list appears on request.
- Alarm and event texts are freely configurable by the substation operator.
- Self-supervision function is continuously activated.

5.6 Other Bay Level Components

In addition to Bay Controllers, other essential relays and devices are needed inside switchyard control rooms to carry out the complete set of SAS functionalities. Some of them are:

- *Connectivity devices*: This includes Ethernet switches and other network components.
- *Protective relays*: This comprises all the set of main and backup relays dedicated to different protections schemes, such as busbar protection, transformer protection and line protection.

Figure 5.7 View of a control cubicle. Source: © Corpoelec. Reproduced with permission of Corpoelec

- *Voltage regulating relay*: Applies when substation equipment includes power transformers provided with on-load tap changers.
- *Parallel control unit*: Applies when two or more power transformers operate in parallel.
- *Auxiliary power system controller*: This controller, similar to the bay controller, in some cases includes the logic for operating low voltage automatic transfers.
- *Fiber-optic distribution panel*: Besides the control network itself, other subsystems may also require fiber-optic distribution, such as disturbance recording systems and metering systems.

5.7 Process Bus

This local area network proposed by the standard IEC 61850 adds another step toward building substation control systems free of copper wiring. It is based on the installation of analog/digital converters (merging units and switchgear drivers) near the primary apparatus, in order to transmit signals to the bay controller by using fiber-optic media. In case of large substations, the bay junction box, already mentioned in Chapter 4, may be a good place to locate such devices.

Although the concept has become controversial due to insertion of additional elements along the control chain that affect availability, the solution is slowly beginning to be implemented by utilities and other substation owners around the world.

Further Reading

Aanestad, H., Deter, O., Rohsler, H., et al. (1998) Substation earthing with special regard to transient ground potential rise – Design aims to reduce associated effects, paper 23–06, *CIGRE session.*

Alexander, S.B. (1997) *Optical Communication Receiver Design*, Spie Press/IEE, Bellingham.

Ari, N. and Blumer, W. (August 1987) Transient electromagnetic fields due to switching operations in electric power systems, *IEEE Trans. Electromagnetic Compatibility*, Vol. EMC-29, pp. 510–515.

Au, M., Gagnon, F. and Agba, B.L. (Aug. 2013) An experimental characterization of substation impulsive noise for a RF channel model, *Progress In Electromagnetic Research Symposium Proceedings, Stockholm.*

Baass, W., Brand, K-P. and Menon, A. (2007) Acceptable function integration of protection and control at bay level, paper 217, *Colloquium of CIGRE Study Committee B5, Madrid.*

CIGRE Task Force of Working Group 33.11 (1999) Flashovers of open circuit-breakers caused by lightning strikes, *Electra Review* 186, 115–121.

CIGRE WG 36.04 (Dec. 1997) Guide on EMC in Power Plants and Substations.

Delaballe, J. (2001) EMC: electromagnetic compatibility, Cahier Technique no. 149, *Technical Collection*, Schneider Electric.

Ghania, S.M. (June 2013) Study of the transient electromagnetic fields inside high voltage substations, *Engineering Science and Technology: An International Journal* 3(3).

Grcev, L. (1996) Transients voltages coupling to shielded cables connected to large substation earthing systems due to lightning, paper 36–201, *CIGRE session.*

Hofbauer, F. (1988) The protection of high voltage substations against lightning, paper 33–02, *CIGRE session.*

IEC 60071–1 Standard, Insulation Co-ordination – Part 1: Definitions, Principles and Rules.

IEC 60071–2 Standard, Insulation Co-ordination – Part 2: Application Guide.

IEC 60099–4 Standard, Surge Arresters – Part 4: Metal-Oxide Surge Arresters Without Gaps for AC Systems.

IEC 60255–1 Standard, Measuring Relays and Protection Equipment – Part 1: Common Requirements.

IEC 60255–26 Standard, Measuring Relays and Protection Equipment – Part 26: Electromagnetic Compatibility Requirements.

IEC 60529 Standard, Degree of Protection Provided by Enclosures (IP Code).

IEC 61312–1 Standard, Protection Against Lightning Electromagnetic Impulse – Part 1: General Principles.

IEC 61578–2 Standard, Mechanical Structures for Electronic Equipments – Tests for IEC 60917 and 60297 – Part 2: Seismic Tests for Cabinets and Racks.

IEEE 1313.2 Standard, Guide for the Application of Insulation Coordination.

IEEE Std. 299, Standard Method for Measuring the Effectiveness of Electromagnetic Shielding Enclosures.

IEEE Std. 693, Recommended Practice for Seismic Design of Substations.

Indulkar, C.S., Kothari, D.P. and Ramalingam, K. (2010) *Power System Transients– A Statistical Approach*, PHI Learning Private Limited, New Delhi.

Johns, A.T. and Salman, S.K. (1995) *Digital Protection for Power Systems*, Peter Peregrinus Ltd/IEE, London.

Jones, B. (1973) Switching surges and air insulation, *Philosophical Transactions of The Royal Society of London, Series A, Mathematical and Physical Sciences* 275(1248), 185–180.

NEMA Standard 250, Enclosures for Electrical Equipment (1000 Volts Maximum).

Ott, H.W. (2009) *Electromagnetic Compatibility Engineering*, John Wiley & Sons, Ltd, Chichester.

Rakov, V.A. and Uman, M.A. (2003) *Lightning: Physics and Effects*, Cambridge University Press.

Rashhes, V.S. and Ziles, L.D. (July 1996) Very high frequency overvoltages at open air EHV substations during disconnect switch operations, *IEEE Transaction on Power Delivery* 11(3).

Schneider Electric Cahier Technique 141, Electrical Disturbances in LV.

Thapar, B., Gerez, V. and Emmanuel, P. (January 1993) Ground resistance of the foot in substation yards, *IEEE Transaction on Power Delivery*, 8(1).

van der Sluis, L. (2001) *Transients in Power Systems*, John Wiley & Sons, Ltd, Chichester.

Varju, G., Szilagyi, F. and Nemeth, J. (June 1999) Mitigation of EMI in high voltage substation environment by use of wiring cables with improved screening effectiveness, *International Conference on Power Systems Transients, Budapest.*

Wiggins, C.M. (October 1994) Transient electromagnetic interference in substations, *IEEE Trans. on Power Delivery* 9(4).

6

Station Level: Facilities and Functions

Station level refers to the place from where the substation is controlled and monitored as a whole. At this level the central substation controller (station controller), the local operating facilities (human machine interface: HMI) and the means for communicate with remote upstream control level (e.g., a network control center) are accommodated. This chapter aims to give an overview of different component details and the tasks they are dedicated to.

6.1 Main Control House

This building, generally located near the switchyard (see an example in Figure 6.1), must be appropriately sized to accommodate the following:

- *Control room*: The place for the HMI mounted at an operator's work station. Ideally, a window should be installed from which the opening and closing processes of disconnectors located in the switchyard can be seen (see Figure 6.2). Inside this room there are usually cubicles containing: the station controller, Ethernet switches, protection system master unit, communication devices, metering devices, the disturbance recording system master unit, fiber-optic rack distributors, fire detection and suppression systems and AC and DC power distribution cubicles.
- *Storage room*: This room is equipped for suitable storage of both heavy components and smaller devices and accessories reserved as spare parts.
- *Office room*: Occasionally, a clear space may be required that is useful for personnel meetings and drawing displays by technical staff.
- *Amenities and toilet facilities*: A separate lunch room and a toilet services room accessible through an internal door.

Substation Automation Systems: Design and Implementation, First Edition. Evelio Padilla.

Figure 6.1 View of a main control house. Source: © Corpoelec. Reproduced with permission of Corpoelec

Figure 6.2 View of a control room. Source: © Corpoelec. Reproduced with permission of Corpoelec

The building must be big enough to accommodate the pertinent SAS components for the ultimate substation stage. The design needs to cover all required provisions for different services such as: Ducts for connection with GPS antenna, cable trays, air conditioning and fire detection and suppression devices.

6.2 Station Controller

The station controller (also called the central substation processor, station computer or front end), is an industrial grade computer running an operative network system , supplemented with additional modules/cards needed to allow the management of essential services like Ethernet connectivity and time synchronization signals. This IED manages and coordinates the functioning of the different bay controllers.

The main process and interfaces related functions of the station controller include the following:

- Communication with bay controllers through the station bus.
- Communication with HMI through the station LAN.
- Storage of substation parameters data.
- Data preparation for communication with the network control center.
- Recording of events with an adequate time resolution (e.g., less than 1 ms).
- Acquiring data related to status and performance of the primary power circuit.
- Providing a time synchronization signal to the bay controllers.
- Adding time stamping to those values that are not time stamped at the bay level.
- Processing signals for the inter-bay interlocking logic.
- Participation in the logic of the bus coupling operation.
- Compilation of alarm and event lists.
- Calculation and storage of active and reactive power on a cyclic basis.
- Immediate communication of any/all abnormal substation conditions to the NCC.
- Releasing printing orders of alarms and events.

In this device, the data (groups of bits representing events, alarms, status indication and digital/analog values) are stored at two different levels. The first is the RAM where the collected data is manipulated to create concentrated data, lists and reports to be sent to the NCC and then to be archived. After the data is sent to the NCC and archived, it is no longer necessary to keep them in the RAM and they are overwritten by incoming new data. The archiving itself is normally done by writing to a file on a hard disk where storage is preserved through periodical back-up procedures (using CD-ROM or similar) in order to store the historical data indefinitely. The application programs, as well as the substation object database, are stored in non-volatile memory (NOVRAM, flash or similar).

To enhance SAS availability, in cases of HV substations it is recommended to duplicate the station controller, configuring them, for example, in a hot/standby fashion with identical and synchronized data bases in both units. By doing that, the units will be serially coupled, but any time only one of them will be assigned to the control task. In this case, a self-monitoring scheme is implemented in such a way that each unit can detect the failure of the other and the control task is automatically passed to the "healthy" unit.

6.3 Human Machine Interface HMI

HMI (also called the workplace for operator or master control center) consists of a set of pieces of hardware (see Figure 6.3) plus a package of applications software.

The hardware usually includes the following:

- Color monitors for display screens showing substation power circuits as well as control and monitoring resources.
- Alphanumeric keyboard, or with function keys for interaction with displayed screens, and a mouse or trackball.
- Printer to produce hardcopies on demand.
- Data logger for continuous printing of event texts in chronological order.

As for the HMI software, full-graphic interfaces and editor programs are typically employed.

The screens (screen dump images shown next), which need to be easily comprehended by the substation operator and are developed in close cooperation between the SAS designer and the substation owner staff, comprise various key stages, such as those in the following sections.

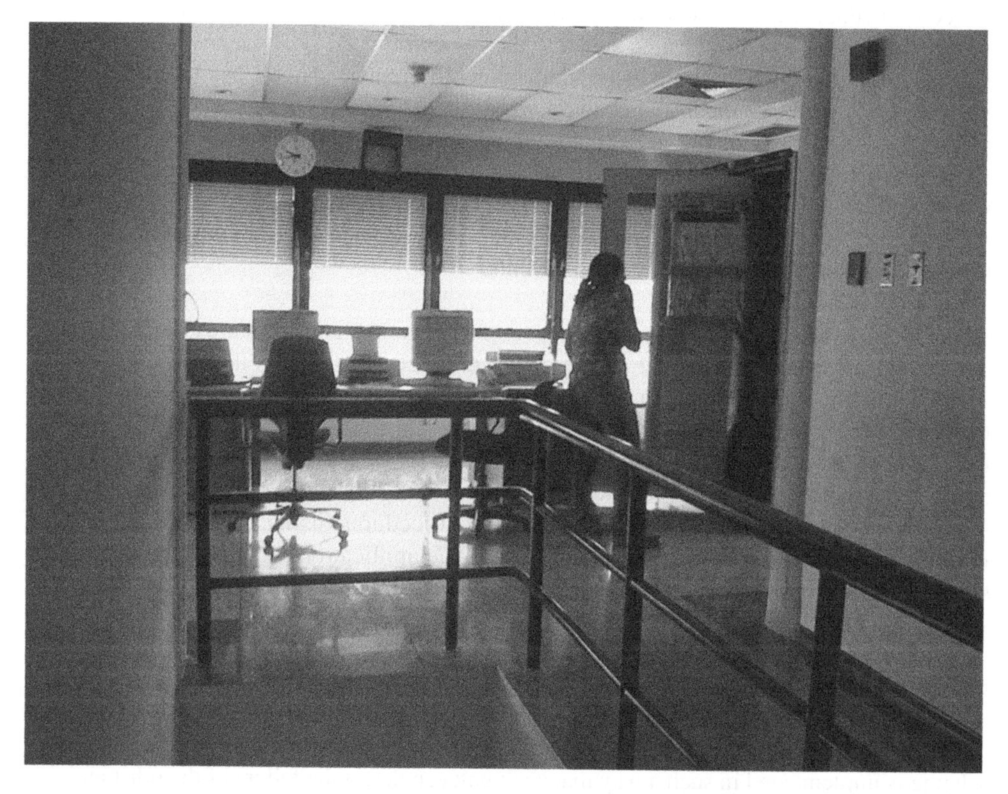

Figure 6.3 View of a HMI. Source: © Corpoelec. Reproduced with permission of Corpoelec

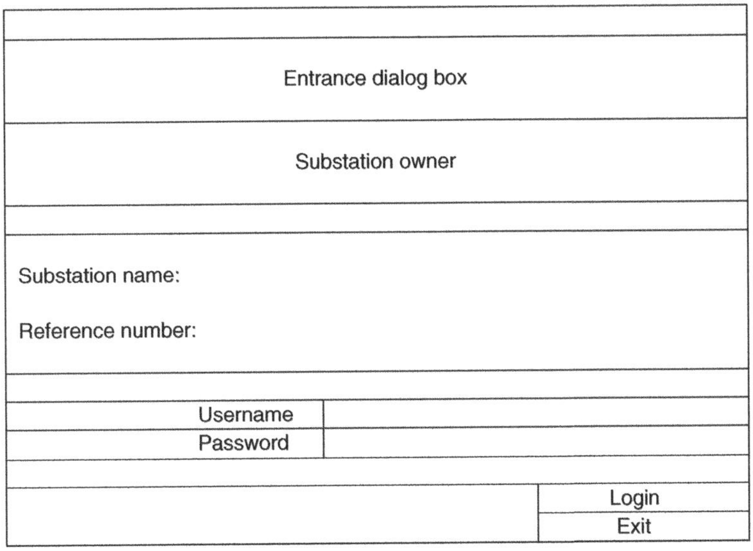

Entrance dialog box

Substation owner

Substation name:

Reference number:

Username	
Password	

Login

Exit

Figure 6.4 Example of the start-up screen

6.3.1 *Start-Up Screen*

It is common that an entrance dialog box, like that shown in Figure 6.4, requests prospective users to enter a username and password for prevention of unauthorized access to the switchgear operation.

6.3.2 *Main Box Screen*

This is the first screen that appears after login through the entrance dialog box. Buttons that appear on this screen include those in the following:

- Fields for data and time information.
- Enter to the event list.
- Enter to the alarm list.
- Exit key.

6.3.3 *Users Administrator Screen*

By clicking on the "Options" button shown on main box screen, several secondary buttons appear giving the option to choose other required screens, including a "users administrator" box such as that in Figure 6.6. Priority levels indicated on this screen are often established according to the following principles:

- Level 1: Allows only an SAS overview and to read different data.
- Level 2: Allows control operations.

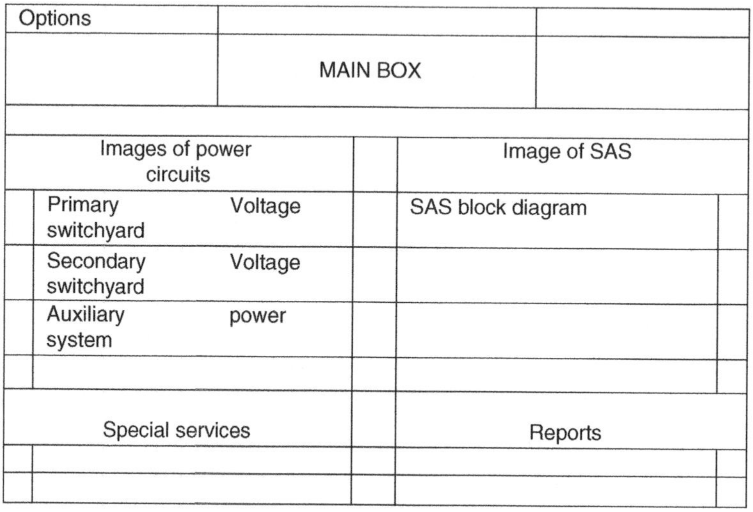

Options

MAIN BOX

Images of power circuits

Image of SAS

Primary switchyard Voltage

SAS block diagram

Secondary switchyard Voltage

Auxiliary power system

Special services

Reports

Figure 6.5 Example of main box screen

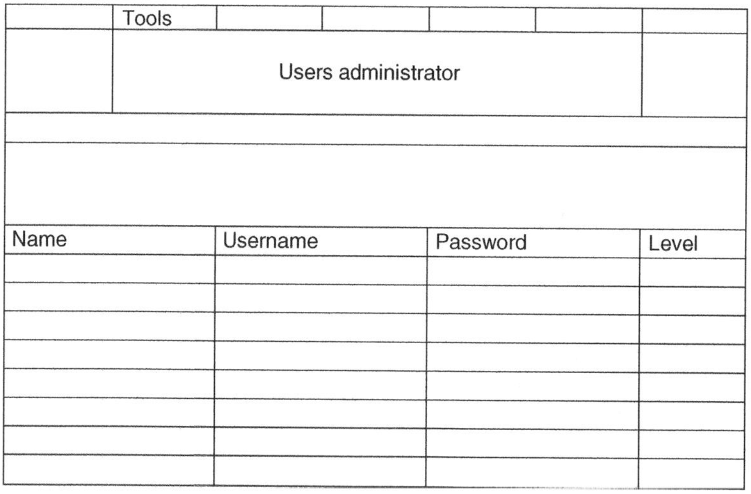

Tools

Users administrator

Name	Username	Password	Level

Figure 6.6 Example of an user administrator screen

- Level 3: Allows the full range of control operations, including modifications of tap changer positions.
- Level 4: Allows engineering tasks, including modification of algorithms and database maintenance work.
- Level 5: Allows all rights, including the authority to add and remove system users.

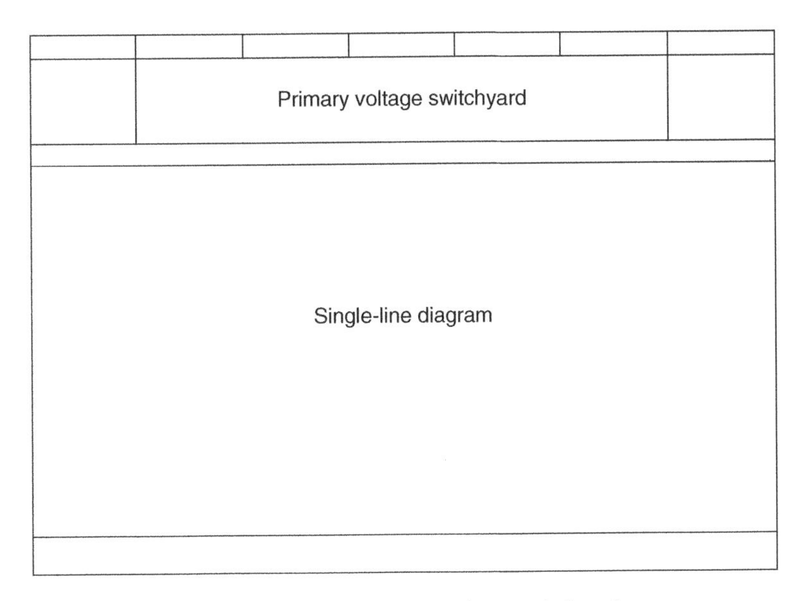

Figure 6.7 Example of primary voltage switchyard screen

6.3.4 *Primary Circuit Screen (Process Screen)*

This screen appears as a result of clicking the respective button on the main box screen (see Figure 6.7). Besides the single-line diagram, identification codes of primary apparatus and names of connected feeders, it indicates the following:

- Positions of local/remote selectors installed on switchgear boards and control IEDs.
- Status of all installed switchgear (open/closed).
- Values of measurements at relevant points of the primary circuit.
- Means to open complementary dialog boxes for particular control functions, like tap changer control.

6.3.5 *SAS Scheme Screen*

This screen appears after clicking on the respective button on the main box screen (see Figure 6.8). In addition to SAS components and network configuration, the screen also indicates:

- Status information of the hardware (abnormal conditions).
- Means to open complementary dialog boxes for control and protection device parameterization.

6.3.6 *Event List Screen*

This screen shows all events related to switchgear or protective relays that are important for substation operation. Special care must be taken during the SAS engineering stage to ensure that events appearing in this list are shown in strict chronological order (see Figure 6.9).

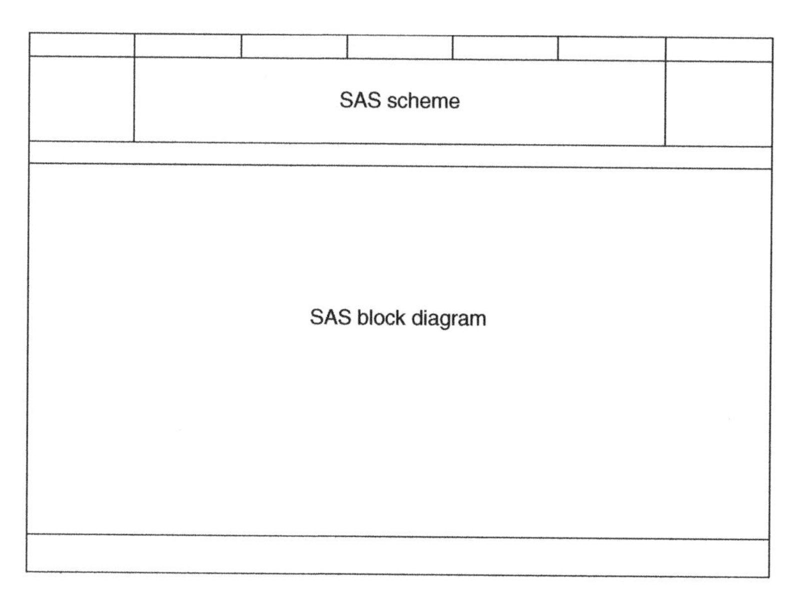

Figure 6.8 Example of a SAS scheme screen

Event list

Date	Time	Apparatus	Event text	

Figure 6.9 Example of an event list screen

6.3.7 Alarm List Screen

This screen provides of features aimed to optimize usefulness for the substation operator, for example:

- Means to indicate if an alarm has been acknowledged by the operator.
- The class of the alarm according to the problem it is showing (see Figure 6.10).
- A scrolling effect.

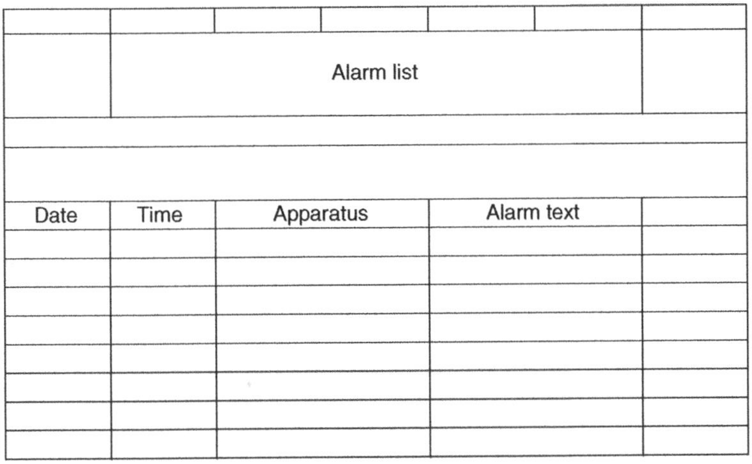

Alarm list

Date	Time	Apparatus	Alarm text	

Figure 6.10 Example of the alarm list screen

As for the station controller, in very important substations, it is also recommended that the HMI facilities are duplicated in case of availability issues on a hot/standby basis. Provision must be made to allow back-up of historical data onto an external or built-in mass storage unit.

The HMI as a whole must be developed according to the latest international ergonomic standards and easy; logical methods must be provided for system start-up and shut-down.

6.4 External Alarming

Alarms are essential help for substation operation. They advise when any component is faulty or when any risk is present to different control levels. This makes alarm display a critical attribute of SASs. Because of that, a lot of particular dedicated alarm annunciator devices are commercially available. The use of one such alarm annunciator as a parallel path for monitoring critical parameters is recommended. The main features of a good alarm annunciator are:

- Self-contained device to get harmonization at the operator desk.
- A minimum of 16 on/off alarm channels.
- Provided with a relay output to activate an audible alarm.
- Comes with a serial communication port for connecting to a station LAN.
- Fully programmable by means of buttons located on the front panel.

Other factors to be considered on this matter are the following:

- *Functional effectiveness*: This refers to the capability of the device for getting the attention of the substation operator during abnormal conditions by use of adequate visual means, such as using super-bright LED indicators.
- *Immunity against electrical interference*: The field contacts used for device sensing in the monitored parameter must be potential-free in such a way that the voltage level needed to activate the alarms is provided by an internal DC power source.

- *Channels programming*: All alarm channels must be individually programmable for operation from normally closed contacts or from normally open contacts.
- *First-out indication*: The alarm annunciator must be provided with a means to indicate the sequence in which alarms are activated on different alarm channels.

6.5 Time Synchronization Facility

In the past, only a few control and monitoring devices were needed with clock synchronization. Nowadays, with the introduction of the IEC 61850 Standard including its process bus solution, all devices connected to the SAS network have to be time synchronized. The time synchronization signal is used to adjust the internal clocks of control IEDs, protective relays, Ethernet switches and other secondary devices. This allows the availability of accurate information needed for making control decisions and for global analysis of power system behavior, especially for understanding the fault escalation process.

As well as time stamping events and alarms, time synchronization is also required by the following applications:

- Data transmission through IEC 61850 communication services applied to the station bus (GOOSE messaging and MMS services).
- Data transfer from merging units to bay controllers by using the process bus concept.
- Data processing by protective relays.
- Fault recording function.

The time synchronization signal may be distributed to SAS components using a direct method or through station bus and process bus networks. The first is only technically feasible when few devices have to be synchronized. In both cases, the signal come from a synchronization time server sourced from a GPS antenna. The accuracy of the signal may fall in the range of micro-seconds or milliseconds, depending mainly on communication protocol features and network traffic load.

6.6 Protocol Conversion Task

Most substations are geographically dispersed parts of power systems and have to operate in coordinated manner under supervision of a common regional or national power Network Control Center (NCC). Because of this, a reliable communication link is used, such that a remote control station located at the NCC collects relevant information under the master/slave principle (switchgear status, events, alarms …) from all supervised substations in real time.

Usually, the remote control station does not manage the same communication codes as the SASs implanted at different substations. This brings about the need to match the communication capability between local and remote control devices. Such a method for this is the protocol converter, popularly known as Gateway.

Traditionally, a protocol converter was a separate device with its own CPU, until few years ago when the option appeared to delegate the protocol conversion task to the station controller CPU. This last solution increases the reliability of the communication link because of fewer physical series-connected elements along the data transmission chain. When a separate device is used, it is often connected to station controller, although in some SAS solutions it is connected to a star coupler device, which is also a good practice, especially if station controller is not duplicated. If the station controller/star coupler are duplicated, an automatic switch for changing the connection to the active unit is required.

6.6.1 Briefing on Digital Communication Protocols

All communication process are based on certain premises and rules that have to be respected and practiced by all that participate in the communication act. A digital communication protocol is a well-structured package of rules and formats for transferring communication messages (datagrams) between networked devices by using streams of binary digits (0s and 1s). Typical communication messages to be transferred in SAS/NCC environments include the following:

- *High priority messages*: Messages belonging to this group generally contain a light digital code (small collection of bits) representing command orders or data for short instructions that have to be transmitted in just a few milliseconds, such as trip, close, block, release and similar signals.
- *Medium priority messages*: Refers to all messages containing data that need time stamping, for example events and alarms data.
- *Low priority messages*: These are messages for getting or sending less time-critical data, such as reading or changing device settings or the winding temperature value of power transformers.
- *Untreated data messages*: This group of messages includes mainly sampled values coming from process bus transducers (merging units) associated with primary current transformers and voltage transformers.
- *File transferring messages*: This type of message applies to large consolidated amounts of data, particularly for recording purposes.
- *Time synchronization messages*: These messages allows transmission of signals from the time-server in order to synchronize the internal clock of different SAS devices.

Digital communication protocols are generally characterized by the following architectural and functional attributes:

- A set of predefined frame formats to accomplish the data transmission process with the required efficiency.
- A mechanism to preserve a high level of data integrity.
- A data fragmentation mechanism to be applied when the message to be transmitted is too large for a single standardized datagram size.
- An error detection mechanism.

6.6.2 *Premises for Developing Protocol Conversion*

It is recommended that early in the SAS design stage, the substation owner and SAS designer agree on how the protocol conversion will be conceived. This is achieved by clarifying key items, such as the following:

- Hardware to be used at the station level:
  - Station controller
  - Stand-alone front-end PC
  - Commercially available devices
- Information regarding the control master remote unit:
  - Supplier
  - Model/version/level
  - Other details
- Protocols to be coupled:
  - Substation side (IEC 61850 platform, other …)
  - Remote terminal side (DNP, IEC 60870–5, other …)
  - Specific versions of protocols on both sides
  - Specific levels to be implemented
- Device profile documents:
  - Substation side (to be submitted by the SAS provider)
  - Remote terminal side (to be submitted by the substation owner)
- Location of remote terminals:
  - NCC 1
  - NCC 2
  - Substation owner headquarters
  - Other locations
- Types of connections:
  - By using modems
  - Point to point connections
  - Other types of connections
- Features of connection devices:
  - Baud rate
  - Other features
- Person responsible for the supply of connection devices (at substation and remote terminal):
  - SAS provider
  - Substation owner
  - Third party
- Responsibility for work on complete data transmission path (installation, test of connections …):
  - SAS provider
  - Substation owner
  - Third party staff
- Type of data to be transferred:
  - Command orders for primary switchgear
  - Command orders for secondary switchgear

    - Measuring values
    - Alarms (individually/grouping)
    - Counter values
    - Sequence of events
    - Command orders for power transformer tap changers
    - Changes in SAS device settings
    - Other types of data
- Hierarchical order between different control levels:
    - First: bay level, second: station level, third: remote terminal level
    - Other hierarchical orders
- Testing facilities (master unit emulator/protocol analyzer):
    - Provided by the SAS designer
    - Provided by the substation owner
    - Provided by others.

6.7 Station Bus

The bay control and station control levels are linked by a high speed communication network called the station bus (also known as the inter-bay bus). The physical structure of this bus consists of a fiber-optical arrangement to which the various upper parts of SAS devices are coupled. Through this bus, the station controller periodically polls all bay controllers ports in a predefined sequence (polling class data); also, the station controller may receive data sent spontaneously by different bay controllers (unsolicited responses).

In a modern SAS, this network operates under the IEC 61850 Platform using MMS (client/server) communication service for the communication between bay level and station level devices (vertical communication) and MMS plus GOOSE services for communication between bay level devices (horizontal communication).

6.8 Station LAN

This less sophisticated communication network is the means to network the majority of station level elements (station controller, HMI, engineering workstation, printer server, etc.). The connection between different devices is usually hardwired but it can be even made by fiber-optic cables.

Further Reading

Apostolov, A. and Monnier, S. (1997) Object-oriented design of human-machine interface for substation integration systems, *Electrical and Computer Engineering: IEEE 1997 Canadian Conference on Engineering Innovation: Voyage of Discovery* (Vol. 2).

Castillo J. (2001) Recent developments in the substations control systems in EDELCA transmission network, *CIGRE SC-23 Colloquium*, Ciudad Guayana.

CIGRE (1997) Database management for tele-control systems, *Technical Brochure* 46.

CIGRE (2000) The use of IP technology in the power utility environment, *Technical Brochure* 153.

Comer, D.E. (2013) *Internetworking with TCP/IP Volume 1*, 6th Edition. Addison-Wesley, Chichester.

Fodero, K., Huntley, C. and Whitehead, D. (2010) Wide-Area Time Synchronization for Protection and Control, *Proceedings of the 36th Annual Western Protective Relay Conference, Spokane*.

Hossenlopp, L., (May/June 2007) Engineering perspectives on IEC 61850, *IEEE Power & Energy Magazine*, pp. 45–50.

IEC 60870-5-104, Telecontrol equipment and systems – Part 5–104: Transmission Protocols – Network access for IEC 60870-5-101 using standard transport profiles.

IEEE Std. 1379, Recommended Practice for Data Communications between Remote Terminal Units and Intelligent Electronic Devices in a Substation.

IEEE Std. 1588, Standard for a Precision Clock Synchronization Protocol for Networked Measurement and Control Systems.

IEEE Std. 979, Guide for Substation Fire Protection.

Lebow, I. (1998) *Understanding Digital Transmission and Recording*, IEEE Press, Piscataway.

Nibbio, N., Genier, M., Brunner, C., et al. (2010) Engineering Approach for the End User in IEC 61850 Applications, *CIGRE Session*.

Pozzuoli, M.P. (n.d.) The need for "Substation Hardener" Ethernet Switches, Paper for RuggedCom Inc – Industrial Strength Networks.

Rim, S-J., Zeng, S-W. and Lee, S-J. (2009) Development of an intelligent station HMI in IEC 61850 based substation, *Journal of Electrical Engineering & Technology* 4(1), 13–18.

Seifert, R. (2008) *The All-New Switch Book: The Complete Guide to LAN Switching*, John Wiley & Sons, Ltd, Chichester.

Udren, E., Kunsman, S. and Dolezilek, D. (April 2000) Significant substation communication standardization developments, *Proceedings of the 2nd Annual Western Power Delivery Automation Conference, Spokane*.

7

System Functionalities

Substation Automation Systems (SASs) have a wide range of responsibilities. These include some extremely critical actions such as clearing faults on time to preserve the physical condition of power system components, to providing more cosmetic facilities, for example an office type remote terminal solely for showing the status of substation primary equipment in a business environment. To accomplish its important mission, the SAS must execute various functions, as explained next.

7.1 Control Function

Control function can be performed from different control levels:

- From bay level: Through the bay controller screen and associated accessories.
- From station level: Using color monitors and other elements from the Human Machine Interface (HMI).
- From remote control level: By means of a master control unit.

Due to practical and technical considerations, these control levels usually do not have the same control autonomy. At bay level (local control rooms), each particular bay controller is able to take care of one or perhaps a pair of HV bays/feeders, while from the HMI located in the main control house, an operator may choose any primary or secondary switchgear to connect or disconnect power circuits. The control capability from the remote master unit is often restricted to opening and closing primary circuit breakers.

Substation Automation Systems: Design and Implementation, First Edition. Evelio Padilla.

In order to allow control function from the bay level, many tasks are assigned to the bay controller. These include the following:

- Supervision of bay related voltages.
- Checking voltage synchronization for the closing operation of circuit breakers.
- Runtime verification for disconnector operations.
- Check for the right control level authority (hierarchical control concept).
- Supervision of pumps, spring systems or other elements of a circuit breaker driving mechanism.
- Supervision of SF6 gas pressure in circuit breaker interrupting chambers.
- Check bay/station interlocking conditions.
- Prevent the progress of double operation commands.
- Verify the due automatic/manual operation mode.
- Verify the applicability of the interlock override operation mode.
- Present pictures (single-line diagram and dialog means) for control purposes.
- Self-supervision including the condition of the voltage supplies.
- Detection of pole discrepancy in circuit breakers
- Delivery of switching commands if favorable safety and operational conditions are fulfilled.

Based on that set of facilities and factors, control function is performed through the local control display of the bay controller, following the control concepts shown in Figure 7.1.

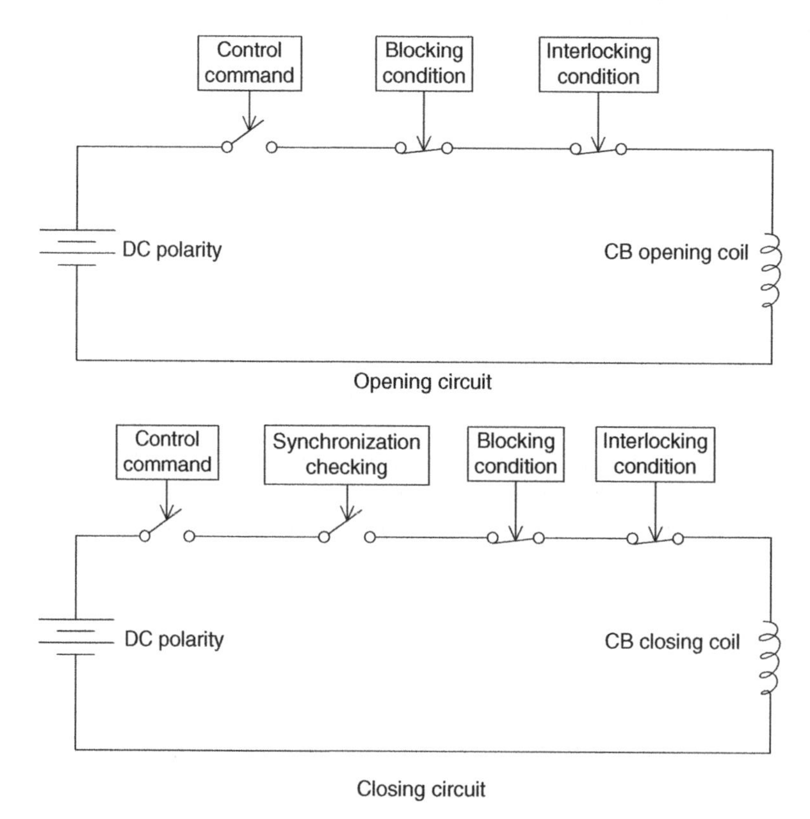

Figure 7.1 Control concepts

At station level, the operator can execute the control function in a similar way to bay level, as there is possibility to select control of any of the primary or secondary switchgear, and also some additional control functions are available reserved for that level; for example, changing positions of on-load tap changers.

7.1.1 Control of Primary Switchgear

As previously mentioned, the primary switchgear consists of HV circuit breakers able to connect/disconnect small or large segments of the power system (feeders, bays, transmission lines) under either normal or abnormal conditions (short circuits or other type of failure), and also of disconnectors needed to isolate such segments of the power circuit, providing the required safety conditions for inspection, maintenance or repair activities.

Circuit breakers, performing these roles, are conformed of sealed interrupting chambers filled with SF6 gas and a powerful operating mechanism for quick operation times (e.g., 50 ms), while disconnectors are conformed of a set of blades moved by a motor-driven operating mechanism, which allows an operation time in the range of a few seconds.

This means that from operational point of view, as well as the common open and closed status that can be exhibited by both apparatus, the disconnector also presents as a third status that referred to in the transit condition exhibited during the operation process.

7.1.1.1 Symbols, Colors and Appearance Representing Primary Switchgear

An efficient visual display on bay controllers and the HMI color monitor is very desirable. This calls for close cooperation between contracting departments at the SAS engineering stage. Although the style for displaying bays and feeders on the screen of the bay controller is often standardized by each IED vendor, a lot of detail may require early agreement between substation owner and SAS designer/integrator. One of these is the set of symbols used to represent primary switchgear under different conditions, so as to reduce the possibility of error in the operative procedure. Some examples of this are shown in Figure 7.2.

7.1.1.2 Switching Command Implementation

On demand, for example by selecting a specific circuit breaker or disconnector on the monitor screen, a control dialog box appears which is used to give the "go to the control" command. In this way, the recommended two-step procedure, consisting of two positive and successive actions to take control, is fulfilled. For security reasons, and following the authority levels rules, adequate means must exist to verify that the person clicking has the correspondent right to execute control commands.

Another important provision to be made is that the internal coupling needs to prevent control commands being made on the same apparatus simultaneously from different control levels, such as NCC and bay level (prevention of double operation commands). The command order has to progress and be materially executed only if no constraint related to interlocking or a blocking condition exists. When there is any impediment carrying out the intended control command, a supplementary dialog box appears indicating the cause of it.

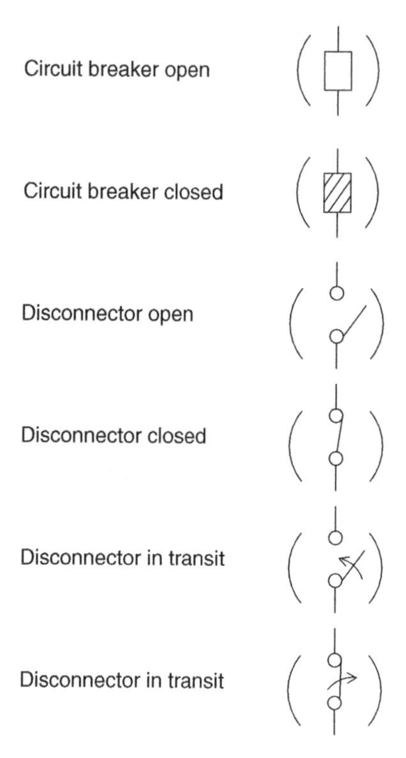

Figure 7.2 Examples of symbols to represent primary switchgear

7.1.1.3 Supervision of Circuit Breaker Trip Circuit

Perhaps the most important SAS functionality is the ability to automatically open specific circuit breakers when a fault occurs in the power system (trip order coming from protective relays). This fact means the trip circuit of circuit breakers is a critical part of the secondary system with respect to its operative availability when it is requested.

Because of that, it is common practice to implement continuous monitoring of such a trip circuit in order to detect defective contacts, weakness in power supplies or any other potential risk. This is done by forcing a current to circulate the circuit and then checking the circuit resistance. When it increases, an alarm appears. For this application, special relays on the market are recommended.

7.1.2 Check of Voltage Synchronization (Synchrocheck)

AC voltage, which operates in most power systems, is a dynamic variable and is characterized by the following parameters:

- *Magnitude*: Usually defined by the RMS value of sinusoidal quantity, for example 115 kV.
- *Frequency*: Refers to the number of sinusoids per second, such as 60 cps.
- *Phase-angle*: A relative parameter indicating time-displacement between different sinusoids. This is the case, for example, in three-phase electric systems, in where the three different phases are equal in magnitude and frequency but displaced by 120° in phase relations.

Table 7.1 Example of matrix for voltage synchronization criteria

Vs1 (live)	Vs2 (live)	Vs1 (dead)	Vs2 (dead)	Vdif Less than Ref-Dif	Fdif Less than Ref-Dif	PAdif Less Than Ref-Dif	Output
true	true	false	false	true	true	true	allow
true	true	false	false	true	true	false	refuse
true	true	false	false	true	false	false	refuse
true	true	false	false	false	false	false	refuse
true	true	false	false	true	false	true	refuse
true	true	false	false	false	true	true	refuse

When a circuit breaker is in the open position, it may have active voltage on one or both sides. When a closing command is given from any control level, the closing action must be effectively performed at the instant when voltages present on both sides of the CB match (with tolerance) with respect to magnitude, frequency and phase-angle. Otherwise, dangerous transient phenomena will occur in the power system.

Because of this, control logic must include certain synchronization criteria, for example those indicated in Table 7.1.

In a modern SAS, the Synchrocheck task is implemented as a function of the bay controller; in the past, a separate device was used instead.

7.1.3 *Checking Operative Constraint*

Like any other type of control system, substation control systems aim to accomplish certain constraint conditions to ensure that all changes in power system configuration are made in a safe manner for personnel, equipment and environment. Those constraint conditions are imposed on the control system by avoiding the progress of a control command when a switch operation is wrong from the operative point of view (interlocking logic) and/or through blocking signals, in cases of conditions unsafe for performing the intended switch operation (blocking condition).

7.1.3.1 Checking of Interlocking Conditions

Several technical reasons impede autonomous switchgear operation. They include the following:

- *Disconnector capabilities*: These apparatus are not able to handle normal load or short-circuit currents when they operate.
- *Earthing switch function*: These apparatus are put in the closed position occasionally for inspection, maintenance or repair activities on a segment of the power circuit. They connect that segment to the earth mesh of the substation. If such a segment is energized when the earthing switch is in the closed position, severe failure will occur (short-circuit).

Because of these factors, the basic interlocking conditions are:

- A circuit breaker shall not close if an associated disconnector is in the transit condition.
- A disconnector shall not open if an associated circuit breaker is in the closed position.
- A circuit breaker shall not close if the associated power circuit is connected to the earth.
- An earthing switch shall not close if voltage is present on the associated power circuit.

Due these restrictions, the control logic of switchgear considers not only its own elements but also external elements from other switchgear in the same bay (bay interlocking), as well as those coming from switchgear installed in different bays in some cases (inter-bay interlocking). Such elements consist basically of indication signals regarding the status (current position) of the respective switchgear.

Although substation owners have conservatively kept the means to exchange those signals between different bays by using hard-wiring based on reliability arguments, a few years ago data exchange between bay controllers using the goose service provided by the IEC 61850 platform started. With this solution, inter-bay interlocking data do not need require any more processing by the station controller.

7.1.3.2 Checking of Blocking Conditions

Switching operations require optimal condition of the respective switchgear. A circuit breaker will be explosive if an opening procedure is made slowly due to an anomaly in the operating mechanism. A disconnector may cause severe damage in instrument transformers if its opening or closing time greatly exceeds typical operation time. These undesirable scenarios bring about the necessity to avoid risking switching operation that could affect substation workers and/or involuntarily interrupt the energy flow. The security in this respect is achieved by implementing blocking signals into the control logic in such way that the switching command (opening/closing) does not progress if any of those blocking signals are active.

Typical blocking signals include:

- SF6 gas – low pressure
- Hydraulic/pneumatic/spring – system defective
- Pole discrepancy in circuit breakers
- Motor faulty.

7.1.4 Voltage Regulation Task

Voltage magnitude is a changing parameter in the power system. It varies with time, is different from a place to other and depends mainly on active and reactive power fluxes along transmission lines. Due to operative reasons, it is essential that voltage magnitude at any point of the power system falls into a predefined acceptable range, for example +/– 5% of the rated voltage. To keep the voltage magnitude inside the acceptable range, a large power system may require periodic connection/disconnection of shunt reactors and capacitor banks, or in some cases have an even more sophisticated installation like Static-Var-Compensators (SVC) permanently active.

In most HV substations power transformers have a motor-operated mechanism called the On-Load Tap Changer (OLTC), which allows a change in the number of turns in the transformer winding by selecting a given number of available taps. By this means, secondary voltage of the power transformer and consequently the voltage on associated power circuit can be adjusted to an acceptable level. OLTCs are controlled (changing tap positions) through a specially designed Voltage Regulator Relay (VRR) installed in the transformer control panel or inside a local control room.

The VRR senses the actual voltage magnitude on the power circuit, compares it with a reference value, and if it is applicable, automatically sends a command order to the tap changer motor. The VRR is coupled to the bay controller associated with the power transformer feeder, in order to allow manual tap changer control from appropriated control levels.

7.1.5 Parallel Working of Power Transformers

HV Power Transformers (three-phase units or three-phase arrangements formed of single-phase units) can be designed to handle up to hundreds of MVAs. However, in some cases it is necessary to connect two or more of them in parallel, in such a way that they share the electric load associated with a certain segment of the power system. Ideally, installing transformers with exactly the same internal characteristics and setting the different VRRs to the same reference voltage would be the solution. In practice, even if the transformers come from the same manufacturer, small differences in internal impedances produce imbalances in parallel operation. This brings about the need to use a specially designed electronic module for working in conjunction with all VRRs. The paralleling module receives status information from circuit breakers and disconnectors to create a schematic image of the actual substation configuration showing which transformers are operating in parallel.

Additionally, the module must also receive the actual values of magnitude and phase-angle of the current from each transformer from VRRs. Based on these values the module calculates the circulating reactive current of each transformer. A control signal derived from such circulating reactive currents is given to one or several VRRs, which control the respective tap changer to the "Raise" or "Lower" tap position until a minimum circulating reactive current is reached at each power transformer.

7.1.6 Operation of Secondary Components

Moreover dealing with primary switchgear, SASs are also useful for controlling secondary components, for example:

- Closing and opening MV circuit breakers.
- Operating LV automatic transfer switches.
- Turning diesel generators on and off.

Medium voltage circuit breakers (usually SF6 gas or vacuum isolated) are used mainly to protect auxiliary power transformers from internal failures. The means to control these apparatus are similar to primary switchgear, from the LCD of an IED located in a local control room and from the HMI at station level.

Automatic transfer switches are dedicated to providing adequate availability levels in power supplies that serve essential loads into the substation. Inputs for the control logic of this device include signals of minimum and maximum voltage levels at incoming feeders, as well as status indications from associated LV circuit breakers.

The diesel generator is a back-up power source in case of total loss of other AC power sources, such as those coming from the tertiary winding of power transformers or from external feeders. The controls to activate and connect that equipment are generally part of an automatic transfer scheme.

7.1.7 Facilities for Operation under Emergency Conditions

This refers to those SAS means specially conceived for performing control actions beyond normal substation operation circumstances. Two such facilities are: A set of accessories to be added to the bay controllers for override interlocking logic. This facility shall be provided with a safety device in order to prevent its use by unauthorized personnel. The second is a highly visible push-button recommended to be installed near the HMI at the main control house. That button is used to open the circuit breakers associated with power transformers only in extreme situations, for example when smoke or fire are present around power transformer.

7.2 Monitoring Function

Knowing what is happening in the controlled process is essential to control function. In HV substations there are many parameters and conditions that merit continuous operator attention. Even if nothing abnormal is occurring in the power system, it is necessary to check, show and record a lot of variables reflecting power system behavior and the operative conditions of substation components. Of course, if any anomalous situation becomes present in the substation, it must be reported immediately to the appropriated control levels.

This is achieved by means of a well-structured group of SAS facilities addressed at maintaining the sending of information to power system actors and to providing opportune warning when any risk factor appears.

7.2.1 Event Handling

Typical substation events include:

- Changes in switchgear status (open/closed)
- Changes in position of local/remote selectors
- Changes in position of manual/automatic selectors
- Trip of MCBs
- Activation of protective relays
- Changes in alarm status (acknowledged, non-acknowledged, blocked)
- Metal enclosed switchgear extracted
- Start/stop of the diesel generator
- Actuation of automatic transfer switches
- Alarm activations
- Failures in switching commands.

Event handling refers to the complete set of artifices implemented at the SAS engineering stage to accomplish the goal of bringing a clear, friendly and reliable event list to the hands of the operator. These include:

- *Time stamping*: Every event must be marked with the time at which it occurs and as close as possible to the place where it is detected.
- *Reliable event collection*: Generally, bay controllers are provided with an event buffer to ensure that no event is lost even in event avalanche scenarios.
- *Event texts*: Each expected event is named by preparing an event list together with the substation owner, in such a way that event names are defined at different control levels in a customized manner and using native language.
- *Event presentation*: The purpose of the event list requests that all events appear on the list in strict chronological order.
- *Storage capability*: Bay controllers need to be equipped with enough memory to store a reasonable minimum amount of events, for example 15,000.

7.2.2 External Disturbance Recording

In some cases, generally for large transmission substations, electrical utilities require a sophisticated Disturbance Recording System to be installed in the substation in order to have enough data and extra facilities to analyze extensive failure and other perturbations. These systems are often configured in a distributed form consisting of several bay units plus a master unit. Although the system works in an independent manner, it is linked to the rest of the SAS mainly to monitor the condition of different components.

7.2.3 Alarming Management

Alarm messages are a critical service provided by SASs. They have to ensure that the operator is aware of any unexpected fact or condition that may require quick intervention. Alarm messages typically relate to the following:

- Circuit breaker pole discrepancies.
- SF6 gas pressure.
- Exceeded run time in disconnector operations.
- Over-voltage conditions.
- Under-voltage conditions.
- Switching blocking conditions.
- Oil levels in power transformers.
- Voltage lost in secondary circuits.
- Over-run of hydraulic pumps.
- Over-load of driving mechanisms.
- Trip circuit failures.
- Temperature values in power transformers.
- Buchholz relay activated.
- Over-pressure in valve operation.

- Transformer cooling system.
- Failures in protective relays.
- Failures in parallel working of power transformers.
- Failures on SAS components.
- Loss of communication at HMI facilities.
- Loss of communication with remote control center.

Alarm management features include:

- *Message texts*: Each alarm message must be named by preparing an alarm list together with the substation owner, in such a way that alarm messages are defined at different control levels in a customized manner.
- *Alarm classification*: It is common practice to differentiate between various alarm messages based on the priority level of the factor that caused it.
- *Alarm grouping*: This is made by certain alarm messages that have a common root, in order to transmit alarm signals to the NCC.
- *Display means*: Corresponds to methods displaying different devices shown in messages or gives an audible signal (LCD, color monitors, external alarm annunciator and loudspeakers).
- *Message presentation*: There are several attributes given to the alarm messages in order efficiently process the perception and handling of information. These include using color combinations and flashing effects.
- *Handling tool*: Covers any means to choose handling options, such as presentation mode, acknowledgement and inhibition possibilities or filtering procedures.

7.3 Protection Function

The protection function is performed by a particular set of IEDs (protective relays) that are connected to the station bus together with bay controllers. These relays monitor voltage and current actual values on primary circuits for continuous comparison with predefined limit values. When actual values drift out of acceptable range, the relay goes into an active condition at the first stage, until a trip-condition is reached if the abnormality persists. Trip orders are addressed in a coordinated manner to open specific circuit breakers, in such a way that the segment of the power system in which the abnormality (failure) occurs becomes isolated from the rest of the system. Based on that segment-oriented principle, typical protection schemes include:

- Busbar protection
- Line protection
- Transformer/shunt reactor/capacitor bank protection.

As the first group of substation IEDs were developed using digital technology a few decades ago, today most protective relays are IEC 61850 compliant.

7.4 Measuring Function

The measuring function is needed to take control decisions and to check the performance of the power system. Control decisions, made by local or remote operator, may comprise:

- Manual voltage regulation.
- Connection/disconnection of shunt reactors and capacitor banks.
- Connection/disconnection of transmission lines.
- Load shedding.
- Energy management tasks.

Regarding power system performance, the measuring function allows the operator be certain about factors like:

- Power flow through different power circuits.
- Load conditions of power transformers.
- Load conditions of transmission lines.

In order to offer such possibilities, based on direct measurement of voltages, currents and phase-angles, the SAS (station controller) is provided with the means to calculate:

- Phase to phase voltage
- Power factor
- Active power
- Reactive power
- Energy.

7.5 Metering Function

Distribution substations are equipped with metering systems. Unlike measuring results, which are useful for operational purposes, quantities displayed by metering devices are used for commercial invoicing. This part of the SAS acts as an independent branch consisting of high precision devices integrated in a dedicated communication network. Special care must be taken to ensure that individual metering devices exhibit adequate features to receive metrological certification by local authorities.

7.6 Report Generation Function

People like information instead of unprocessed data. The SAS collects a large quantity of data coming from bay controllers, protective relays, digital fault recorders and other IEDs installed in the substation. Although part of that data is used for real time signals through various SAS screens such as those related to the single line diagram and event and alarm lists, the data can be processed, grouped and presented in the form of reports not only useful to operation personnel but also for various utility staff teams. Some examples of such reports are:

- Peak demand of active power as a time function.
- Temperature values in power transformers as a time function.

The reports can be presented as full graphic curves on a two-dimensional coordinate system, as well as in a tabular mode. Periodicity may also be assigned, such as creation of daily, weekly or monthly reports.

7.7 Device Parameterization Function

Several IEDs that form part of the SAS, particularly protective relays, are designed in such a way that they have adjustable parameters, for example the activation current value in an over current relay. Facilities of modern SASs include dialog boxes for fixing or changing (settings) such parameters.

Further Reading

Bevrani, H., Watanave, M. and Mitani, Y. (2014) *Power System Monitoring and Control*, Wiley-IEEE Press, Chichester.

Bosma, A. and Thureson, P-O. (November 2001) A new reliable operating mechanism for HV AC circuit-breakers, *IEEE/PES T&D Conference*.

Cadick, J., Capelli-Schellpfeffer, M. and Neitzel, D. (2006) *Electrical Safety Handbook*, 3rd Edition, McGraw-Hill, New York.

CIGRE Working Group 39.04 (August 1999) Measurements of Quality in Electric Systems, *Electra Review* 185, 115–127.

CIGRE (August 2000) User Guide for the application and diagnostic techniques for switching equipment for rated voltage of 72.5 kV and above, *Technical Brochure* 167.

CIGRE Working Group 13.04 (February 1999) Shunt capacitor bank switching – stresses and test methods, *Electra Review* 182, 165–189.

Cukalevski, N. and Jones, H. (December 1999) Power system operator – training program design, development and utilization, *Electra Review* 187, 117–129.

Fulchiron, D. (July 1998) Protection of MV/LV substation transformers, Groupe Schneider *Cahier Technique* 192.

Horowitz, S.H., Phadke, A.G. and Niemira, J.K. (2014) *Power System Relaying*, 4th Edition, John Wiley & Sons, Ltd, Chichester.

Hux, G. (1984) The influence of the operating mechanism on the reliability of the high voltage circuit breaker, Paper 3, Session III, *International Seminar on Switchgear and Controlgear,* Bombay.

IEC 60214-1, Tap-Changers – Part 1: Performance Requirements and Test Methods.

IEC 60214-2, Tap-Changers – Part 2: Application Guide.

IEC 60617, Graphical Symbols for Diagrams.

IEEE Std. 18, Standard for Shunt Power Capacitors.

Kosakada, M., Watanabe, H., Ito, T., et al. (2002) Integration substation systems – harmonizing primary equipment with control and protection systems, *IEEE/PES T&D Conference and Exhibition Asia Pacific*.

Schneider-Electric (March 1996) Automatic transferring of power supplies in HV and LV networks, *Cahier Technique* 161.

IEEE Std. C37.20.3, Standard for metal-enclosed interrupter switchgear.

Wache, M. (2006) Switchgear interlocking with IEC61850, *China International Conference on Electricity Distribution*.

8

System Inputs and Outputs

Like any other control and monitoring system, the Substation Automation System (SAS) receives a quantity of input signals from both primary equipment and secondary components. These input signals, which are mainly binary, are processed to carry out system functions including the stellar task of changing the power system configuration when necessary. Such control functions are performed through output signals delivered by SAS devices consisting mainly of opening and closing commands addressed to primary and secondary switchgear. To allow the reader to appreciate the physical significance of mentioned input and output signals, a summary of their related details follows.

8.1 Signals Associated with Primary Equipment

Substation primary equipment includes switchgear, instrument transformers and power transformers.

8.1.1 Switchgear

Primary switchgear covers circuit breakers, disconnectors and earthing switches. Associated signals are related mainly to status indication and condition monitoring of operating mechanisms as shown next.

8.1.1.1 Signals Associated with Circuit Breakers

As already mentioned in Chapter 4, HV circuit breakers are powerful apparatus able to interrupt electric currents circulating through primary circuit under both normal load conditions (e.g., 2000 A) and fault conditions (e.g., 30,000 A). Due to that capability, CBs execute the

Substation Automation Systems: Design and Implementation, First Edition. Evelio Padilla.

critical mission of connecting and disconnecting electrically different segments belonging to the power system such as bays/feeders, power transformers, shunt reactors, capacitor banks and transmission lines.

Circuit breakers are conformed of one or several sealed interrupting chambers in which a dielectric gas is contained (mainly SF6 gas), an insulating structure often based on porcelain insulators, a set of elements for providing the mechanical power needed to close and open the primary contacts quickly and a series of devices required for handle control commands and sense relevant parameters and conditions.

Due the demanding service to which the CB is dedicated, strict monitoring and signaling must be made to ensure that the equipment will respond in an appropriate manner when a switching operation is requested from protective relays or a bay controller. This brings about the need to manage in the SAS projects; a number of input signals referring to CB position indication and condition monitoring of various supporting media, as well as output signals consisting of opening/closing command orders, such as those shown in Table 8.1.

8.1.1.2 Signals Associated with Disconnectors

HV disconnectors are simply designed apparatus that allow isolation of power system segments for inspection/maintenance purposes. Typically, a motor-operated mechanism moves a set of conducting blades creating an air gap that separates certain parts of the primary circuit from the rest of the power system.

Input signals coming from disconnectors refer mainly to the position of the equipment (open, closed, transit), while output signals are addressed to transmit opening and closing command orders, like indicated in Table 8.2.

8.1.1.3 Signals Associated with Earthing Switches

This apparatus, generally operated manually, consists of a set of conducting blades for occasional connection to the earth mesh specific segments of a primary circuit. Input signals coming from the equipment informs the SAS of its position (open or closed), such as indicated in Table 8.3.

8.1.2 Instrument Transformers

Instrument transformers include voltage transformers and current transformers. Associated signals basically consist of actual values of voltages and currents present in their secondary circuits.

8.1.2.1 Signals Associated with Voltage Transformers

Voltage transformers allow transfers of a sample of the actual voltage present on primary circuit to the SAS. Additional input signals inform the SAS when a MCB dedicated to protect a secondary circuit is tripped, as shown in Table 8.4.

Table 8.1 Signals associated with circuit breakers

Origin of the signal	Sensor	Interface element	Signal signification	IED destination	Purpose
DC supply Trip-coil 1	Under-voltage relay	Contact	DC voltage lost	BC	Alarm
DC supply Trip-coil 2	Under-voltage relay	Contact	DC voltage lost	BC	Alarm
Operating mechanism (per pole if applicable)	Auxiliary relay	Contact	Position indication (open/closed)	BC	Signalizing/Event/Interlocking
Control cubicle	Local/remote selector	Contact	Position indication (Local/Remote)	BC	Signalizing/Event/control hierarchy
Control cubicle	Auxiliary relays	Contact	Pole discrepancy	BC	Event/Trip
SF6 gas system (per pole if applicable)	Pressure Switch	Contact 1° stage	Gas pressure low	BC	Alarm
SF6 gas system (per pole if applicable)	Pressure Switch	Contact 2° stage	Gas pressure excessively low	BC	Block
Operating mechanism (per pole if applicable)	Pressure Switch (if applicable)	Contact 1° stage	Oil pressure low	BC	Alarm
Operating mechanism (per pole if applicable)	Pressure Switch (if applicable)	Contact 2° stage	Oil pressure excessively low	BC	Block
Operating mechanism (per pole if applicable)	Failure detector (if applicable)	Contact	Hydraulic circuit faulty	BC	Alarm
Operating mechanism (per pole if applicable)	Relay (if applicable)	Contact	Hydraulic pump over run	BC	Alarm
Operating mechanism (per pole if applicable)	Relay (if applicable)	Contact	Hydraulic pump over load	BC	Alarm
Operating mechanism (per pole if applicable)	MCB	Contact	Motor MCB tripped	BC	Alarm
–	–	–	Opening command coil 1 (per pole if applicable)	CB	Open the CB
–	–	–	Opening command coil 2 (per pole if applicable)	CB	Open the CB
–	–	–	Closing command (per pole if applicable)	CB	Close the CB

Table 8.2 Signals associated with disconnectors

Origin of the signal	Sensor	Interface element	Signal signification	IED destination	Purpose
Operating mechanism (per pole if applicable)	Auxiliary relay	Contact Type 1*	Position indication (open)	BC	Signalizing/event/ interlocking
Operating mechanism (per pole if applicable)	Auxiliary relay	Contact Type 2*	Position indication (transit/closed)	BC	Signalizing/event/ interlocking
Operating mechanism (per pole if applicable)	Auxiliary relay	Contact Type 3*	Position indication (closed)	BC	Signalizing/event/ interlocking
Operating mechanism (per pole if applicable)	Auxiliary relay	Contact Type 4*	Position indication (open/transit)	BC	Signalizing/event/ interlocking
Control cubicle	Local/ remote selector	Contact	Position indication (local/remote)	BC	Signalizing/event/ control hierarchy
Operating mechanism (per pole if applicable)	Failure detector	Contact	Driver failure	BC	Alarm
–	–	–	Opening command (per pole if applicable)	DI	Open the DI
–	–	–	Closing command (per pole if applicable)	DI	Close the DI

Key: *Contact types

Contact Type	DI open	DI in transit	DI closed
Type 1	———		
Type 2		———	———
Type 3			———
Type 4	———	———	

Table 8.3 Signals associated with earthing switches

Origin of the signal	Sensor	Interface element	Signal signification	IED destination	Purpose
Operating mechanism (per pole if applicable)	Auxiliary relay	Contact	Position indication (open/closed)	BC	Signalizing/event/ interlocking

Table 8.4 Signals associated with voltage transformers

Origin of the signal	Sensor	Interface element	Signal signification	IED destination	Purpose
Secondary box	Winding 1	Voltage circuit	Actual voltage	BC/PR	Signalizing/ measurement/metering
Secondary box	Winding 2	Voltage circuit	Actual voltage	BC/PR	Signalizing/ measurement/metering
Bay junction box	MCB	Contact	Winding 1 MCB trip	BC	Alarm
Bay junction box	MCB	Contact	Winding 2 MCB trip	BC	Alarm

Table 8.5 Signals associated with current transformers

Origin of the signal	Sensor	Interface element	Signal signification	IED destination	Purpose
Secondary box	Winding 1	Current circuit	Actual current	BC/PR	Signalizing/measurement
Secondary box	Winding 2	Current circuit	Actual current	BC/PR	Signalizing/measurement
Secondary box	Winding 3	Current circuit	Actual current	BC/PR	Signalizing/measurement
Secondary box	Winding 4	Current circuit	Actual current	BC/PR	Signalizing/measurement/ metering

8.1.2.2 Signals Associated with Current Transformers

The input signal coming from current transformers carries a sample of the actual current flowing through the primary circuit to be processed by protective relays and bay controllers to the SAS, as indicated in Table 8.5.

8.1.3 Power Transformers

Most of the input signals associated with power transformers are to protect their internal components, which are sensitive to temperature values, as well as to detect abnormal conditions present in transformer devices and subsystems. Associated output signals are control orders for a change in position of the on-load tap changer, such as indicated in Table 8.6.

8.2 Signals Associated with the Auxiliary Power System

Equipment belonging to auxiliary power system includes medium and low voltage circuit breakers, medium voltage transformers, distribution cubicles and other components. The associated signals are designed mainly to preserve the integrity of offered services and the physical condition of different system components.

8.2.1 Signals Associated with MV Circuit Breakers

Input signals coming from this equipment refer mainly to position indication, while output signals transmit opening/closing control commands. See Table 8.7.

Table 8.6 Signals associated with power transformers

Origin of the signal	Sensor	Interface element	Signal signification	IED destination	Purpose
AC supply	Under-voltage relay	Contact	Lost AC voltage	BC	Alarm
DC supply	Under-voltage relay	Contact	Lost DC voltage	BC	Alarm
Cooler system, main AC power supply	MCB	Contact	Position indication	BC	Event
Cooler system, emergency AC power supply	MCB	Contact	Position indication	BC	Event
Cooler system, group 1	MCB	Contact	Cooler group faulty	BC	Alarm
Cooler system, group 2	MCB	Contact	Cooler group faulty	BC	Alarm
Tap changer	Driver system component	Contact	Position indication	Voltage Regulator Relay	Signalizing/tap changer control
Tap changer	Over-current relay	Contact	Overload in tap-changer circuit	BC	Alarm
Tap changer	Fault indicator	Contact	Tap changer faulty	BC	Alarm
Tap changer	Supervisor relay	Contact	Tap changer faulty	BC	Trip
Oil system	Flow valve	Contact	Excessive oil flow	BC	Alarm
Oil system	Oil level sensor	Contact	Oil level low	BC	Alarm
Oil system	Oil temperature monitor	Contact	Oil temperature high	BC	Alarm
Oil system	Oil temperature monitor	Contact	Oil temperature excessive	BC	Trip
Oil system	Buchholz relay	Contact	Gas generation	BC	Alarm
Oil system	Buchholz relay	Contact	Gas generation Excessive	BC	Trip
Oil system	Pressure release valve	Contact	Internal over-pressure	BC	Trip
Tap changer oil system	Oil level sensor	Contact	Oil level low	BC	Alarm
HV winding	Winding temperature system	Contact	HV winding high temperature	BC	Alarm
HV winding	Winding temperature system	Contact	HV winding high temperature	BC	Trip
MV winding	Winding temperature system	Contact	MV winding high temperature	BC	Alarm

Table 8.6 (*continued*)

Origin of the signal	Sensor	Interface element	Signal signification	IED destination	Purpose
MV winding	Winding temperature system	Contact	MV winding high temperature	BC	Trip
HV bushing	Bushing type TC	Current circuit	Actual current in the HV winding	Voltage Regulator Relay	Work in parallel of transformers
HV bushing	Bushing type TC	Current circuit	Actual current in the HV winding	Protective Relay	Differential protection
MV bushing	Bushing potential device (if exist)	Voltage circuit	Actual voltage at MV terminal	Voltage regulator Relay	Voltage regulation
MV bushing	Bushing potential device (if exist)	Contact	BPD faulty	BC	Alarm
Tank	Leakage current relay	Contact	Leakage current present	BC	Alarm
	–	–	Rise command	Tap changer	Rise tap position
	–	–	Lower command	Tap changer	Lower tap position

Table 8.7 Signals associated with MV circuit breakers

Origin of the signal	Sensor	Interface element	Signal signification	IED destination	Purpose
Operating mechanism	Auxiliary relay	Contact	Position indication (open/closed)	BC-AS	Signalizing/event/ interlocking
Control cubicle	Local/remote selector	Contact	Position indication (local/remote)	BC-AS	Signalizing/event/ control hierarchy
–	–	–	Opening command	MV CB	Open the CB
–	–	–	Closing command	MV CB	Close the CB

8.2.2 Signals Associated with MV Distribution Transformers

Input signals used to preserve the condition of MV distribution transformers are indicated in Table 8.8.

8.2.3 Signals Associated with LV Circuit Breakers

Input signals coming from LV circuit breakers are generally related to position indication of both the equipment itself and its L/R selector for control. Output signals transmit control orders generated by protective relays or bay controllers (see Table 8.9).

Table 8.8 Signals associated with MV distribution transformers

Origin of the signal	Sensor	Interface element	Signal signification	IED destination	Purpose
Oil system	Buchholz relay	Contact	Gas generation	BC-AS	Alarm
Oil system	Buchholz relay	Contact	Gas generation Excessive	BC-AS	Trip
Oil system	Pressure release valve	Contact	Internal over-pressure	BC-AS	Trip
Oil system	Temperature detector	Contact	High temperature	BC-AS	Alarm
Oil system	Temperature detector	Contact	Too high temperature	BC-AS	Trip CB on secondary side
Oil system	Temperature detector	Contact	Excessive temperature	BC-AS	Trip CB on primary side

Table 8.9 Signals associated with LV circuit breakers

Origin of the signal	Sensor	Interface element	Signal signification	IED destination	Purpose
Operating mechanism	Auxiliary relay	Contact	Position indication (open/closed)	BC-AS	Signalizing/event/ automatic transfer
Control cubicle	Local/remote selector	Contact	Position indication (local/remote)	BC-AS	Signalizing/event/ control hierarchy
–	–	–	Opening command	LV CB	Open the CB
–	–	–	Closing command	LV CB	Close the CB

8.2.4 Signals Associated with Distribution Center "A"

Distribution centers receive electric energy through main incoming circuits in order to distribute it via a number of downstream feeders. Associated input signals are used for the control logic of transfer devices, which select the incoming circuit to be connected to the busbars of the cubicle. They also inform when a LV circuit breaker is tripped, such as is summarized in Table 8.10.

8.2.5 Signals Associated with Distribution Center "B"

Input signals coming from Distribution Center "B" are indicated in Table 8.11.

8.2.6 Signals Associated with AC Distribution Cubicles for Essential Loads

AC distribution cubicles are sourced from their respective distribution center. Associated input signals are shown in Table 8.12.

Table 8.10 Signals associated with Distribution Center "A"

Origin of the signal	Sensor	Interface element	Signal signification	IED destination	Purpose
Feeder 1	Voltage relay	Contact	Minimum voltage	BC-AS	Automatic transfer operation
Feeder 1	Voltage relay	Contact	Maximum voltage	BC-AS	Automatic transfer operation
Feeder 2	Voltage relay	Contact	Minimum voltage	BC-AS	Automatic transfer operation
Feeder 2	Voltage relay	Contact	Maximum voltage	BC-AS	Automatic transfer operation
Automatic transfer	Operation mode selector	Contact	Automatic transfer in manual/automatic mode	BC-AS	Automatic transfer operative
Downstream feeders	LV CBs	Contacts	LV CB tripped	BC-AS	Alarm

Table 8.11 Signals associated with Distribution Center "B"

Origin of the signal	Sensor	Interface element	Signal signification	IED destination	Purpose
Feeder 1	Voltage relay	Contact	Minimum voltage	BC-AS	Automatic transfer operation
Feeder 1	Voltage relay	Contact	Maximum voltage	BC-AS	Automatic transfer operation
Feeder 2	Voltage relay	Contact	Minimum voltage	BC-AS	Automatic transfer operation
Feeder 2	Voltage relay	Contact	Maximum voltage	BC-AS	Automatic transfer operation
Automatic transfer	Operation mode selector	Contact	Automatic transfer in manual/automatic mode	BC-AS	Automatic transfer operative
Downstream feeders	LV CBs	Contacts	LV CB tripped	BC-AS	Alarm

Table 8.12 Signals associated with AC distribution cubicles for essential loads

Origin of the signal	Sensor	Interface element	Signal signification	IED destination	Purpose
Feeder 1	Voltage relay	Contact	Minimum voltage	BC-AS	Automatic transfer operation
Feeder 1	Voltage relay	Contact	Maximum voltage	BC-AS	Automatic transfer operation
Feeder 2	Voltage relay	Contact	Minimum voltage	BC-AS	Automatic transfer operation
Feeder 2	Voltage relay	Contact	Maximum voltage	BC-AS	Automatic transfer operation
Automatic transfer	Operation mode selector	Contact	Automatic transfer in manual/automatic mode	BC-AS	Automatic transfer operative
Downstream feeders	MCBs	Contacts	MCB tripped	BC-AS	Alarm

8.2.7 *Signals Associated with Diesel Generators*

Diesel generators are equipped of a series of subsystems needed for correct operation. Input signals coming from this equipment (Table 8.13) are mainly addressed at monitoring and signalizing of parameters belonging to such subsystems. Associate output signals serve as the on/off switch to the Generator.

8.2.8 *Signals Associated with AC Distribution Cubicles for Nonessential Loads*

Input signals associated with this cubicle are summarized in Table 8.14.

Table 8.13 Signals associated with the diesel generator

Origin of the signal	Sensor	Interface element	Signal signification	IED destination	Purpose
Control cubicle	Local/remote selector	Contact	DG in Local/Remote operation mode	BC-AS	Signalizing/event/control hierarchy
Oil system	Temperature detector	Contact	Temperature high	BC-AS	Alarm
Oil system	Temperature detector	Contact	Temperature too high	BC-AS	Trip
Oil system	Pressure detector	Contact	Pressure low	BC-AS	Alarm
Oil system	Pressure detector	Contact	Pressure too low	BC-AS	Trip
Water system	Temperature detector	Contact	Temperature high	BC-AS	Alarm
Water system	Temperature detector	Contact	Temperature too high	BC-AS	Trip
Mechanical system	Speed detector	Contact	Over-speed	BC-AS	Trip
Winding system	Over-current relay	Contact	Over-current	BC-AS	Trip
Control cubicle	Failure detector	Contact	DG defectives	BC-AS	Alarm
Diesel tank	Level indicator	Contact	Diesel level low	BC-AS	Alarm
Battery	Voltage relay	Contact	Low voltage in battery	BC-AS	Alarm
–	–	–	Turn-on command	DG	Start the DG
			Turn-off command	DG	Stop the DG

Table 8.14 Signals associated with AC distribution cubicles for nonessential loads

Origin of the signal	Sensor	Interface element	Signal signification	IED destination	Purpose
Upstream feeder	Voltage relay	Contact	Under- voltage	BC-AS	Alarm
Upstream feeder	Voltage relay	Contact	Over-voltage	BC-AS	Alarm
Upstream feeder	LV CB	Contact	LV CB tripped	BC-AS	Alarm
Downstream feeder	MCBs	Contacts	MCB tripped	BC-AS	Alarm

8.2.9 Signals Associated with DC Transfer Switches

Input signals coming from the DC transfer switch (Table 8.15) are used for alarming purpose and also for device operation.

8.2.10 Signals Associated with DC Distribution Cubicles

Input signals coming from DC distribution cubicles are shown in Table 8.16.

8.2.11 Signals Associated with Each Voltage Level of Batteries and Chargers

A lot of input signals come from batteries and chargers (see Table 8.17) mainly for the purposes of monitoring and alarming.

Table 8.15 Signals associated with DC transfer switches

Origin of the signal	Sensor	Interface element	Signal signification	IED destination	Purpose
Busbar 1	Voltage relay	Contact	Minimum Voltage	BC-AS	Automatic Transfer operation
Upstream feeder busbar 1	LV CB	Contact	LV CB tripped	BC-AS	Alarm
Busbar 2	Voltage relay	Contact	Minimum Voltage	BC-AS	Automatic Transfer operation
Upstream feeder busbar 2	LV CB	Contact	LV CB tripped	BC-AS	Alarm
Busbar 1	Earth fault relay	Contact	Earth fault in busbar 1	BC-AS	Alarm
Busbar 2	Earth fault relay	Contact	Earth fault in busbar 2	BC-AS	Alarm
Automatic transfer	Operation mode selector	Contact	Automatic transfer in manual/automatic mode	BC-AS	Automatic transfer operation
Downstream feeder	MCBs	Contacts	MCB tripped	BC-AS	Alarm

Table 8.16 Signals associated with DC distribution cubicles

Origin of the signal	Sensor	Interface element	Signal signification	IED destination	Purpose
Busbar 1	Voltage relay	Contact	Under-voltage	BC-AS	Alarm
Busbar 2	Voltage relay	Contact	Under-voltage	BC-AS	Alarm
Upstream feeder busbar 1	LV CB	Contact	LV CB tripped	BC-AS	Alarm
Upstream feeder busbar 2	LV CB	Contact	LV CB tripped	BC-AS	Alarm
Downstream feeder busbar 1	MCBs	Contacts	MCB tripped	BC-AS	Alarm
Downstream feeder busbar 2	MCBs	Contacts	MCB tripped	BC-AS	Alarm

Table 8.17 Signals associated with batteries and chargers

Origin of the signal	Sensor	Interface element	Signal signification	IED destination	Purpose
Battery 1	Earth fault relay	Contact	Earth fault in battery 1	BC-AS	Alarm
Battery 2	Earth fault relay	Contact	Earth fault in battery 2	BC-AS	Alarm
Battery 1	Fuse accessory	Contact	Fuse battery 1 tripped	BC-AS	Alarm
Battery 2	Fuse accessory	Contact	Fuse battery 2 tripped	BC-AS	Alarm
Battery 1	Voltage relay	Contact	Voltage at battery 1 low	BC-AS	Alarm
Battery 2	Voltage relay	Contact	Voltage at battery 2 low	BC-AS	Alarm
Battery room	Fault detector at ventilation system	Contact	Ventilation system faulty	BC-AS	Alarm
Charger 1	Failure detector	Contact	Charger 1 faulty	BC-AS	Alarm
Charger 2	Failure detector	Contact	Charger 2 faulty	BC-AS	Alarm
Charger 3	Failure detector	Contact	Charger 3 faulty	BC-AS	Alarm
Charger 4	Failure detector	Contact	Charger 4 faulty	BC-AS	Alarm
Charger 1	LV CB	Contact	CB of charger 1 tripped	BC-AS	Alarm
Charger 2	LV CB	Contact	CB of charger 2 tripped	BC-AS	Alarm
Charger 3	LV CB	Contact	CB of charger 3 tripped	BC-AS	Alarm
Charger 4	LV CB	Contact	CB of Charger 4 tripped	BC-AS	Alarm

Table 8.18 Signals associated with collateral systems

Origin of the signal	Sensor	Interface element	Signal signification	IED destination	Purpose
Disturbance recorder System	Failure detector	Contact	Disturbance recorder faulty	BC-AS	Alarm
Air conditioned main control room	Failure detector	Contact	Air conditioned defective	BC-AS	Alarm
Air conditioned local control room 1	Failure detector	Contact	Air conditioned defective	BC-AS	Alarm
Air conditioned local control room 2	Failure detector	Contact	Air conditioned defective	BC-AS	Alarm
Air conditioned local control room 3	Failure detector	Contact	Air conditioned defective	BC-AS	Alarm
Fire detection system	Relay	Contact	Fire detection system activated	BC-AS	Alarm

8.3 Signals Associated with Collateral Systems

This refers to those input signals coming from external systems (Table 8.18) that also need to be monitored to keep certain substation environments comfortable and safe, such as those indicated in Table 8.18.

9

System Engineering

The development of the engineering process is crucial in any SAS project. Sometimes it takes longer than expected and in extreme scenarios may put corporate deadlines for accomplishing entire substation projects at risk. The main factors that can detrimentally affect the engineering process include the following:

- Uncertain scope and/or technical requirements in the user specification.
- Prolonged discussions to agree a definitive solution.
- Delays in submitting information.
- Late changes in the scope or functionalities.
- System weakness discovered late.

The content of this chapter intends to help all SAS project actors to carry out the engineering process in an efficient manner.

9.1 Overall System Engineering

Having an early, clear overview of the entire prospective system is a key factor in the SAS engineering process. This allows knowledge in advance about the scope of the work to be developed, it is the guide to define what information is needed from both sides and forms the basis for planning different engineering activities.

This step usually concludes with drawing up a detailed and definitive physical topology scheme in which different real devices and the integrated communication network are represented (instead of just a simplified logic topology scheme).

Substation Automation Systems: Design and Implementation, First Edition. Evelio Padilla.

9.1.1 System General Concept

When SAS projects are contracted in a competition style commercial process, different bidders will offer their simplest, standardized and cheap solutions. Sometimes, the proposed solution comes from a quick interpretation of the user specification written in a foreign language and often such a proposed solution does not include enough information for extensive evaluation by the substation owner.

Once the contract is signed, a clarification stage begins in which the SAS vendor may be surprised by the full set of user expectations, while the user can be disappointed because of the extra cost claimed by the vendor. In such an embarrassing scenario, contracting parties must cooperate with objectivity to agree upon an optimized solution from both the technical and economical points of view, instead of looking for a "perfect solution". At this early stage, typical issues to deal with are the following:

- Number of feeders to be controlled from each bay controller.
- Relationship between control and protection functions.
- Definitive topology.
- Function integration.
- Communication platform.
- Hardware capabilities and functions.
- Application of state-of-the-art solutions.
- List of engineering documents.

9.1.2 System Topology

An elementary principle in SAS projects is that the substation owner must provide all prospective bidders with the freedom to propose their internally standardized topology, which based on previous experience, can be considered to be best suited to the specific case. Similarly, the substation owner should check if the proposed topology fulfills its specification.

In complement to formal availability criteria (e.g., n-1 criteria) as well as to specified redundancy requirements, there is a group of critical SAS capabilities recommended to consider in the process of checking whether the proposed topology may be acceptable. These capabilities include the following:

- No loss of already acquired but not yet treated data at all control levels.
- Possibility for continued and safe manual control (either from remote terminal, station level or bay level) respecting all the interlocking conditions and procedures.
- Possibility of continuing system operation without endangering personnel and equipment (certain degradation in the operation comfort such as automatic operation may be acceptable).
- Capability to keep the remote control safe from the NCC, even if the substation is unmanned.
- Capability to make the time stamping task and storage of events continuous until they can be transmitted to higher hierarchical levels.
- Possibility of maintenance and replacement of defective components without the need to shut down the system.

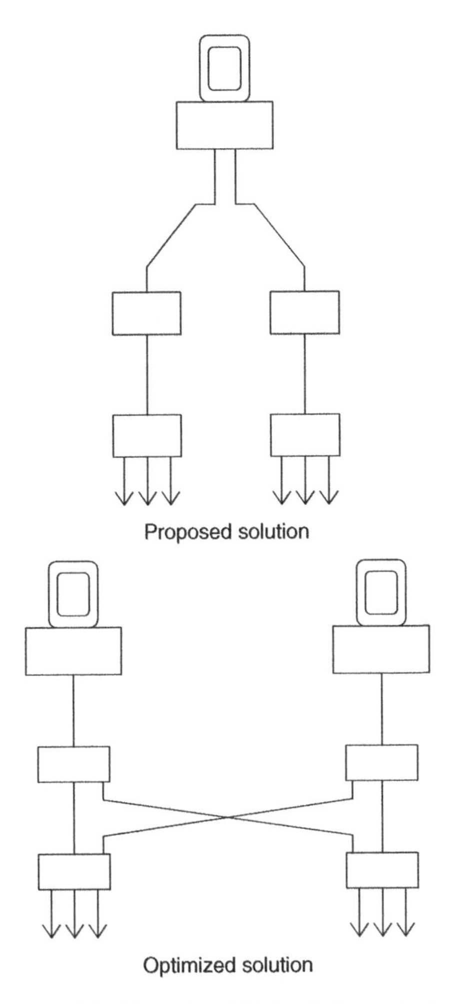

Figure 9.1 Example of SAS solution evolution

The evaluation of the proposed solution may result in small or significant changes to conform the already mentioned optimized solution. An example is shown in Figure 9.1.

9.1.3 Opportune Clarifications

The scheme of system topology does not necessarily say all that substation owners need to know; rather, in some cases it may create confusion or misunderstanding. Because of that, besides the topology, before starting the detailed system engineering, a lot of non-visible issues have to be clarified between vendor and buyer in order to avoid inopportune and polemic discussions. These issues include:

- *System performance*: Refers to relevant operative features under both normal or emergency conditions, such as:
  - Function allocation in different devices.
  - Communication process.
  - Operation mode of duplicated components.
  - Operation possibilities during emergencies.
  - Time stamping procedures.
  - Power consumptions.
- *Limits of the scope of supply*: Interfaces are always gray zones. In cases of SAS projects, particular care must be taken to define early on the party responsible for providing certain facilities, such as A/D converter modules and interface connections for temperatures and tap changer position of power transformers (if SAS vendor or transformer manufacturer). It is recommended that SAS vendors are responsible for providing the complete set of devices and accessories belonging to the communication path to the NCC, including the modem and accessories needed at a remote terminal.
- *Control and protection integration*: The degree of integration must be clear. A common practice is to keep the protection function autonomous for a HV substation and integrated control and protection functions for voltage levels at and below 36 kV.
- *Segregation of IEDs*: Traditionally, electrical utilities have a culture of standardization. In that sense, it is recommended to evaluate the advantages and disadvantage of grouping the bay controllers and protective relays to be connected to a common node, for example to a star coupler device. Segregation criteria may be, for example, based on primary voltage levels or by main and backup units in case of protective relays.
- *Interface with primary equipment*: Even if a hard-wired means was planned in the user specification, exploration of the benefits of using the IEC 61850 process bus is recommended.
- *Redundancies validation*: In some cases, any device may have duplicated power supplies but this is arranged in a supplementary manner. It may induce a false perception of redundancy.
- *Redundancies possibilities*: Further to power supplies, some devices can support redundancies in other active components like the bus master and input/output cards. That possibility may be explored especially for more critical devices, such as star coupler units.
- *Inter-bay interlocking logic*: Current technology allows the implementation of this interlocking category through horizontal communication (bay controller–bay controller) by using the GOOSE service defined by Standard IEC 61850. Although until now only a few practice applications have been reported, it is expected that such means will be commonly applied in the medium term.
- *Measuring allocation*: Most IED manufacturers have integrated the measuring and storage in digital form of currents, voltages, active and reactive power values in their protective relays. This is because these values are necessary for performing the protection function and values can be stored easily without any significant extra CPU loading. This method eliminates the need for cabling the current circuit to the bay controllers, which simplifies the design of the BC bringing the benefits of better security against electromagnetic interference and also leading to a reduction of the overall system cost. In addition, where the protection is duplicated, there are two sets of indications available thus providing redundancy at no additional cost.

- *Disturbance recording*: By evaluating the attributes of the disturbance recording functionality generally integrated into protective relays, it is possible to confirm if any intended separate disturbance recording system are still justified.
- *Protocol conversion*: This task may be performed by a separate device or as a function of the station controller. The preferred solution must be fixed.
- *Communication path to NCC*: It may be the case that an end device, which appears be a protocol converter, is in reality a passive component, for example a router or a switch. This happens particularly in cases of redundant station computers.
- *Hardware data-sheet*: Both parts (buyer and vendor) need to know the strengths and weakness of different system components. Special care is required to quality check those important components coming from third parties, for example Ethernet switches.
- *Engineering tools*: The information to be exchanged between contracting parties when the engineering process starts includes a list of the engineering tools the vendor will use in the particular SAS project.
- *Tests with remote control center*: If these tests are foreseen, it is recommended to make a particular point to clarify key related issues like the methodology to be followed and the provider of master station emulator and traffic monitor device.
- *Cubicle identifications*: A standardized practice in engineering process consists of establishing early on a series of codes for cubicle identification depending on its functions and the location of the associated primary bay. The IEC Standard 81346–1 is recommended to define these identification codes.

9.1.4 *Premises for Engineering Work*

Before starting the engineering work, the substation owner and vendor/integrator must all agree on a set of symbols, colors and conventions to be used in the particular SAS project. These include:

- *Equipment identification*: All switchgear will have a unique identification code based preferentially on user internal standards, for example H2105. Usually, the first character is common to all equipment installed at the same primary voltage level.
- *Switchgear symbols and colors*: Due to switchgear are changing state equipment, a particular form to represent the actual state on the HMI must be agreed. A way to do that is by filling out a form like that shown in Table 9.1.
- *Colors for primary circuits*: It is common that utilities distinguish different primary voltages levels by using different colors when drawing the respective single-line diagrams. Those standardized colors can be agreed filling put a form as shown in Table 9.2.
- *Conditions of process objects*: The process objects to be loaded in the SAS databases (indications, tasks, devices), may exhibit various conditions based on certain pre-defined criteria, such as:
  - *Normal condition*: When a preset value falls into an acceptable range.
  - *Alarm in no acknowledged condition*: When a preset value falls out of an acceptable range or a device or subsystem presents an operative problem. The object shows the attributes to call the attention of the substation operator.

Table 9.1 Example of fixing switchgear position/condition indication

Apparatus	Status	Symbol	Color	Appearance
Circuit Breaker	Open			
Circuit Breaker	Closed			
Circuit Breaker	In manual mode			
Disconnector	Open			
Disconnector	Closed			
Disconnector	Transit			
Disconnector	In manual mode			
Earthing switch	Open			
Earthing switch	Closed			

Table 9.2 Example of form for fixing colors for primary circuits

Primary voltage level	Color
36 kV	
115 kV	
230 kV	
400 kV	

  - *Alarm in acknowledged condition*: When a preset value falls out of an acceptable range or a device or subsystem presents an operative problem. The object keeps the signs of abnormal condition but without calling the attention of the substation operator.
  - *Selected condition*: When a device has been marked by the substation operator to carry out a switching procedure.
- *Tagging modes*: Refers to the facilities give to the substation operator to add advertences or restrictive conditions temporarily on HMI displays. These include the following:
  - *Manual input*: This predefined tag is to be used for shown information unsupervised by the system.
  - *Operator notes*: This is an open type tag to add comments and instructions.
  - *Control inhibited*: When this tag is attached to a specific apparatus symbol, the equipment cannot receive control commands.
  - *Alarm inhibited*: Adding this tag to any process object under alarm conditions means the alarm signal will disappear but the warning of an abnormal condition is still present.
  - *Apparatus in maintenance*: This tag will block the remote operation of the equipment marked.
- *Attributes of process objects according to condition or tagging*: The way process objects should be shown on the HMI under different circumstances must be agreed on, for example filling out a form as in Table 9.3.
- *Attributes of analog values*: The way analog values should be shown on the HMI under different circumstances must be agreed, for example filling a form out as in Table 9.4.
- *Attributes of alarms*: The way alarms should be shown on the HMI under different circumstances must be agreed, for example filling out a form as in Table 9.5.

Table 9.3 Example of form for fixing attributes of process objects

Condition/Tag	Attributes		
	Mode	Color	Appearance
Selected			
Alarm in none acknowledged condition			
Alarm in acknowledged condition			
Control inhibited			
Maintenance condition			

Table 9.4 Example of a form for fixing analogue values attributes

State of the value	Condition	Attributes	
		Color	Mode
Normal	acknowledged		
Normal	none acknowledged		
Abnormal first limit	acknowledged		
Abnormal first limit	none acknowledged		
Abnormal second limit	acknowledged		
Abnormal second limit	none acknowledged		

Table 9.5 Example of format for fixing alarm attributes

Alarm coming from	Color
CB alarm activated	
CB alarm return to normality	
DI alarm activated	
DI alarm return to normality	
Protection alarm activated	
Protection alarm return to normality	
Auxiliary services alarm activated	
Auxiliary services alarm return to normality	

9.1.5 Signals Lists

Although signals lists have some detractors, they are needed to check that all pertinent signals are considered in the engineering scope and also to agree the alarm and event texts to be displayed at different control levels.

9.1.5.1 Signals List Related to the Bay Controller

This list contains the signals and their proprieties, as well as the specific point in where they are received by the IED, as is shown in Table 9.6.

9.1.5.2 Signals List Related to Bay Controller of the Auxiliary Power System

This is similar to the standard bay controller but refers to the auxiliary system.

9.1.5.3 Signals List Related to the Station Controller

This list covers the position indication of all primary and secondary switchgears.

9.1.5.4 Signals List for Communication with the NCC

This list includes all the object information needed at remote control level.

9.1.5.5 Point to Point Signals List (For Each Bay)

This list shows how signals are displayed along all hierarchical control levels, as shown in Table 9.10.

Table 9.6 Example of the bay controller signal list

IED reference		Signal proprieties			Signal name	Schematic diagram	Page
I/O Card	Position	Type	Related object	Status			
03	12	BIN-I	H2105	Open	Circuit breaker position	90-C457	34
03	13	BIN-I	H2105	Closed	Circuit breaker position	90-C457	34
03	21	BIN-O	H2105	Command execute	Closing command	90-C457	34
04	05						

Table 9.7 Example of the BC auxiliary system signal list

IED reference		Signal proprieties			Signal name	Schematic diagram	Page
I/O Card	Position	Type	Related object	Status			
01	12	BIN-I	S1020	Open	MV CB position	17-C457	21
01	13	BIN-I	S1020	Closed	MV CB position	17-C457	21
03	21	BIN-O	S1020	Command execute	Closing command	17-C457	21

Table 9.8 Example of the station controller signal list

Bay identification	Related object	Signal name	Bit-combination			
			00	01	10	11
B4	H2105	Circuit breaker position		open	Closed	
B4	H1105	Disconnector position	Transit	open	Closed	

Table 9.9 Example of the remote control level signal list

Address	Object	Signal name	Acronym
001			
002			
003	H2105	Circuit breaker position	Open/closed
004	H2105	L/R Selector	Local/remote
005			
006			
007			

Table 9.10 Example of a point to point signals list

Equipment (Object)	Bay Level		Station Level			NCC Level			
	Signal	Text	Object	Signal name	Event text	Address	Object	Signal	Status
H2105									

9.1.5.6 Signals Lists Related to Equipment and Systems

Samples of signals lists related to equipment and systems are included in Appendix A.

9.2 Bay Level Engineering

The SAS engineering process at the bay level is essentially an integration process in which the following main elements are involved:

- AC and DC power supplies
- Primary equipment
- Bay controllers
- Voltage regulator relay (if applicable)
- Line protection relay (if applicable)
- Busbar protection relay (if applicable).

The core of engineering work at this level consists of drawing the schematic diagram (one per controlled bay/feeder) showing the following issues:

- AC distribution circuits
- DC polarities distribution
- Switchgear control circuits
- Circuit breaker trip circuits
- Circuit breaker reclosing scheme (if applicable)
- Supervision of trip circuits
- Tap changer control circuit
- Interlocking schemes
- Position indication circuits
- Alarms and blocking circuits.

That work brings the following benefits as a result:

- Checking of the right interconnection between equipment and devices to accomplish the various assigned functions and tasks.
- Determine the composition of bay controller, with respect to the amounts of I/O cards and analog input modules (including reserve inputs)
- Define the application of the alarm LEDs usually located at the front side of the bay controller.
- Provide a means for future troubleshooting by maintenance personnel.

Beside the bay schematic diagram, control cubicle schematic diagrams are also needed. Their contents include:

- Cubicle dimensions
- Layout of components
- List of components
- Terminal identifications
- Front view
- AC power distribution
- DC power distribution.

Other engineering activities at the bay level include the following:

- Configuration of the Synchrocheck functionality into bay controllers.
- Defining the connection of voltage regulator relay.
- Realizing bay interlocking into bay controllers.
- Defining the alarms to be implemented on a separate alarm annunciator.

9.3 Station Level Engineering

Beyond choosing computers and peripheral devices offering a comfortable control desk, strong engineering efforts are required to accomplish the goal of a well-designed SAS. Perhaps the majority of this effort needs be dedicated to the station level of the system.

9.3.1 *Engineering Related to the Station Controller*

The station controller contains a significant amount of software and applications. Although part of those resources come from standardized packages from vendor know-how, each SAS project needs a particular development, which is made adding a lot of new data and programs to get the requested functionalities and facilities.

9.3.1.1 Definition and Implementation of the Station Level Database (Process Database)

Although Substation Automation Systems (SASs) are conceived as distributed intelligence systems, in the station controller resides a central object-oriented database called the process database, useful to perform station-oriented tasks that cannot be done at bay levels. That database is usually structured following a format like the simplified model shown in Table 9.11.

Examples of physical location:
Bay 1, Bay 2, Bay 3, Feeder 1, Feeder 2 …

Examples of equipment/devices/parameter:
CB H2105, CB H2205, CB H2305 …
DI H2110, DI H2115, DI H2120 …
Local/Remote selector,
MCB 1, MCB2, MCB 3, …
Bay Controller B1, Bay Controller B2 …
Line Protection Relay (main), Line Protection Relay (backup) …
Voltage
Current
Frequency
Active power

Table 9.11 Simplified model of a process database

Object origin		Object name	Object status
Physical location	Equipment/device/ parameter/task		

Examples of object names and status for control:

Object name	*Object status*
Opening command	Off/selected
Closing command	Off/selected
Command execution	Off/executed/faulty
Rise tap changer	Off/executed
Lower tap changer	Off/executed
Command mode	Automatic/manual

Examples of object names and status for signaling:

Object name	*Object status*
Position indication	Closed/open
Local/remote selector	Remote/local
SF6 gas pressure	Normal/alarm
Battery condition	Normal/alarm
Oil temperature	Normal/alarm

Examples of object names and status for protection:

Object name	*Object status*
Line protection L1	Normal/started/tripped
Busbar protection B1	Normal/started/tripped
Earth fault protection	Normal/tripped
Buchholz relay second stage	Normal/tripped

9.3.1.2 Implementation of Redundant Solutions

In systems where extremely high availability is required, a redundant solution for the station controller such as a hot/stand-by solution may be applied. That concept is based on data shadowing between two physically separate units in such a way that in case of hot condition unit failure, a switchover takes place meaning that the unit receiving the shadowed data in stand-by mode becomes "hot". Compared with a single unit base solution, for the hot stand-bay solution considerations must be made regarding the high requirements of computer resources such as computing power and memory. Special care must be taken to ensure that only the active unit can exchange data externally.

Engineering considerations also include the analysis of system behavior when the switchover process between both units takes place. In particular, whether or not a loss event will occur.

9.3.2 Engineering Related to the Human Machine Interface

The Human Machine Interface (HMI) is like the "face" of the SAS. It gives the substation operator access to control means as well as alarms and events displayed on the monitor screen.

Besides the alarm list presented on the screen, critical alarms are often also displayed by a separate alarm annunciator and the events are printed out on a dedicated event logger.

The engineering process related to the HMI covers a series of issues to make complete operation and monitoring of the substation comfortable from that place. These issues include those in the following sections.

9.3.2.1 General Design Principles

Users of SAS expect that HMI features are built in an easily comprehensible manner, such that personnel with minimum background knowledge in IT are able to operate the system. A well designed set of HMI displays is characterized by consistency regarding formats, labels, colors and fonts.

A key factor is showing switchgear symbols in single-line diagrams using different colors dependent on the actual condition, for example:

- Normal condition
- Selected for control command
- Alarm state
- Blocked state
- Blocked control
- Faulty state
- Maintenance condition.

9.3.2.2 Typical Screens

The set of screens to be designed during the engineering stage generally includes the following:

- *First entrance menu*: This screen shows buttons accessing different display sections, such as single-line diagrams, measuring results or system topology and components.
- *Entrance dialog box*: This box contains windows to type the username and password belonging to the intended operator.
- *User's configuration box*: This box contains windows to load user names and authority levels.
- *Single-line diagram of primary circuit*: This screen displays the status of the primary equipment in terms of actual values of voltages, currents, frequency, active and reactive powers, as well as the position of circuit breakers, disconnectors, earthing switches and transformer tap changers.
- *Single-line diagram of auxiliary power system*: This screen shows medium and low voltage apparatus and other relevant components like transfer switches and the diesel generator.
- *Control dialog box*: This box is automatically displayed when a circuit breaker or disconnector is selected for control. It is recommended to view this box as a collateral window to the single-line diagram instead of a complete standalone screen picture. Using this box implements the standard control principle of select-before-operate. If the selected switchgear has any impediment to operation, such as an interlocking or blocking condition, the system will show a message and the intended control command is canceled. Once the control command is executed, the switchgear must be shown in the flashing condition until the apparatus has reached its new position.

- *Tap changer control box*: This box, also collateral to the single-line diagram, shows the reference voltage, the tap changer actual operation mode and the means to rise and lower the tap position.
- *Alarm list*: The alarm texts shall be established in a customized way; that is, using an autochthonous SAS user lexicon.
- *Event list*: Restriction in the number of characters to describe the events can be acceptable in favor of maintaining just one line per event in the screen and printout presentation.
- *Measurement dialog box*: Buttons to select specific desirable measuring have to be available, as well as windows to show measured values.
- *System overview*: This screen shows the system's physical topology and components including bay controllers, protective relays, networking devices, HMI components and interfaces and devices for communication with the remote control center. Clicking on a device makes device identification appear and proprieties like the manufacturer, model and version of base software.
- *Relay setting box*: This box will appear when a relay is selected on the system overview screen. It shows the relay proprieties, as well as windows to change relay parameters.

9.3.2.3 Operative Features

The engineering process at HMI level also covers the definition and establishment of several operative issues, such as:

- *System supervision*: This refers to defining adequate mechanisms for obtain a self-monitored system such that faults are immediately flagged to the substation operator before they develop into serious situations.
- *Backup facilities*: Provision must be made at the HMI level to create a backup of directories or files contained into the system.

9.4 Functionalities Engineering

This part of the engineering process covers the develop of all algorithms and programming engaged with fulfilling system functionality requested by the substation owner, as well as the implementation of techniques and solutions addressed to accomplish the best current industry practices.

9.4.1 *Interlocking Engineering*

Software interlocking developed during the SAS engineering stage is addressed to prevent inadvertent incorrect switchgear operation, which can cause damage to substation equipment or injury to personnel. The software is created in a bay-oriented format based on a set of particular rules depending on the primary circuit arrangement of the substation.

In the case of a double busbar scheme, the rules are the following:

1. Tripping by protective relays on all circuit breakers is unrestricted.
2. The opening command on all circuit breakers, other than the busbar coupler circuit breaker is unrestricted.
3. An opening command cannot be given on the busbar coupler circuit breaker when:
   - Both busbar disconnectors are closed in any feeder.
   - Any busbar disconnector or sectionalizing disconnector is in transit condition.

4. A closing command cannot be given on a circuit breaker when its associated line disconnector or busbar disconnectors are in the transit condition.
5. A line disconnector can only be operated when its associated circuit breaker and feeder earthing switch are open.
6. A feeder earthing switch can only be operated when its associated line disconnector, if applicable, and busbar disconnector are open.
7. A busbar disconnector can be operated when its circuit breaker, busbar earthing switch, feeder earthing switch(es) and the other busbar disconnector are all open (except the case indicated on item 8).
8. If a busbar disconnector is closed, the second busbar disconnector can only be operated when the busbar sections concerned are coupled through the busbar coupler circuit breaker.
9. A busbar earthing switch can only be operated when the busbar disconnectors and the busbar sectionalizing disconnector, on the associated busbar, are open.
10. A busbar sectionalizing disconnector can only be operated when all the busbar disconnectors and the busbar earthing switch located on the same side of this sectionalizing disconnector, are open.

9.4.2 Voltage Regulation Engineering

During the engineering stage the following themes are defined:

- From where the input will be taken to be used as the reference voltage for the voltage regulator relay.
- The specific CT:current ratio to be chosen for the purpose of parallel working transformers (if applicable).
- The necessity to apply conversion to BCD code for tap position indication.

9.4.3 Protection Engineering

Many values, most of them related to associated transmission lines, are established in the engineering stage to be used in the process of protective relay parameterization.

9.4.4 Metering Engineering

At the very least, a drawing must be prepared in the engineering stage showing metering devices, connection interfaces, networking devices, and data acquisition units to be located into the substation and at the remote control center.

9.4.5 Disturbance Recording Engineering

If a separate disturbance recording system is installed, a set of signals to start such a system are defined during engineering stage. These are usually the following signals coming from different lines and transformer bays:

- Primary protection trip.
- Backup protection trip.
- Trip orders transferred from remote substations.

9.4.6 System Self-Monitoring Engineering

One of the main advantages of SASs based on digital technology is the capability of self-monitoring. With this in mind, engineering activities cover the means to check the following matters belonging to the system itself.

- Auxiliary supply voltages.
- Availability of the various assemblies.
- Transmission and execution of control commands.
- Data traffic.
- Software behavior.
- Memories performance under different system scenarios including starting and re-initialization.
- Timing periods.
- Consistency between actual switchgear status and the respective displayed status.

9.5 Auxiliary Power System Engineering

The auxiliary power system must be not underestimated in the engineering SAS context, because it has its own complexity and a significant amount of signals to be processed. For controlling this system at the bay level, a bay controller is used similar to that for HV bay application. The upstream control of medium and low voltage switchgear is often restricted to the substation HMI without remote control levels being allowed.

9.5.1 Design Concept

The electrical supply for the auxiliary power system may come from different sources including tertiary windings of power transformers, external medium voltage lines and standby power generators. Dealing with the design concept of auxiliary power system in the engineering stage confirms the definitive scheme to be implemented to get an adequate reliability level. The main issues to be evaluated are capacity and expected availability of the different available power sources.

9.5.2 AC Voltage Distribution

The engineering efforts regarding AC voltage distribution are addressed at defining distribution cubicles with respect to capacity and location, as well as to determine AC circuits at different voltage levels to be installed to serve all substation loads, such as:

- Main control house
- Local control rooms
- Switchyard (junction boxes)
- Power transformers (cooler system, control panel)
- Circuit breakers (motors, pumps)
- Disconnectors (motors).

Detailed studies are necessary to calculate cabling capacities and establish capability and discrimination sequence of low voltage protection devices.

9.5.3 DC Voltage Distribution

The study of DC voltage requirements is a particularly important item of SAS engineering process. It must cover not only the load values but also their coincidence factors under different possible but reasonable scenarios. The criticism of this study is due to its impact on dimensioning cabling systems, protective devices and batteries. In terms of distribution scheme itself, some perhaps re-evaluable criteria are the following:

- The use of separate DC cubicles by power and control disciplines.
- The use of single or double busbar inside DC cubicles.
- The arrangements of DC circuits for serve disperse loads like CB motors (ring or dedicated feeder).
- The options to provide two separate DC power supplies for control and protection devices.

9.5.4 Batteries and Chargers

High quality is the main concern in choosing the batteries. Another important factor is the construction of the rack in which the batteries are to be installed. A low profile structure is recommended to avoid undesirable effects that may be caused by earthquakes. Modern battery chargers are provided with digital control modules whose features and interfaces must be well understood by substation owner.

9.5.5 Medium Voltage Switchgear

A schematic diagram of the medium voltage switchgear must be elaborated at the engineering stage. Its content shall cover at least the following issues:

- Three-phase scheme of power circuit.
- Control and monitoring circuits.
- Protection circuits.
- Lighting and heating circuit.
- AC and DC voltage distribution.
- Trip circuit.
- Closing and opening circuits.
- Signaling circuits.
- Mechanical interlocking system.

9.5.6 Automatic Transfer Switches

Automatic transfer switches are key components of the auxiliary power system. They enable essential loads to be supplied selectively from normal feeders or from alternate sources. The auxiliary power system may have one or several automatic transfer schemes. A careful analysis is required to establish the operational logic of those schemes, which is usually documented by using and/or gate symbols.

9.6 Project Drawings List

A proposed drawing list for a typical SAS project is included in Appendix B.

9.7 The SAS Engineering Process from the Standard IEC 61850 Perspective

The engineering process of the SAS is outlined by the standard based on the use of a set of system integration tools that deal with various files created with the Substation Configuration Language (SCL) defined in IEC 61850–6.

Those tools are the following:

- *The System Specification tool*: This tool is used to define the substation structure from conventional single-line diagram and requested SAS functionalities by creating the System Specification Description file (SSD) considered crucial to reach interoperability.
- *The System Configuration tool*: This tool is able to import the SSD file and the files of device models (ICD) to configure the communication functionalities and create the System Configuration Description file (SCD).
- *The Device Configuration tool*: This tool, applied to configure bay controllers, protective relays and other IEDs, imports the SCD file, creates the device model files and creates the device parameterization files to be loaded into the target device.

As may be perceived, the engineering process as treated by the Standard has no conflict with traditional SAS engineering processes. Because of this, engineering scope and criteria contained in this chapter are also valid when the SAS is fully based on the Standard IEC 61850.

Further Reading

ANSI/IEEE C37.90, Standard for Relays and Relay Systems Associated with Electric Power Apparatus.

Apostolov, A. (April 25–26, 2005) IEC 61850 and disturbance recording, 2005 *Georgia Tech Fault and Disturbance Analysis Conference*.

CIGRE (April 1997) Database management for telecontrol systems, *Technical Brochure* 46.

CIGRE (April 2000) The use of IP technology in the power utility environment, *Technical Brochure* 152.

CIGRE (2001) System protection schemes in power networks, *Technical Brochure* 187.

CIGRE (2002) Design guidelines for power station auxiliaries and distribution systems, *Technical Brochure* 197.

CIGRE Working Group B3-01 (June 2004) Impact of new functionalities on substation design, *Electra Review* 214, 63–75.

IEC 61082, Preparation of Documents used in Electrotechnology.

IEC 61175, Designation of Signals.

IEC 61355, Classification and Designation of Documents.

IEC 61666, Identification of Terminals within a System.

IEC 62023, Structuring of Technical Information and Documentation.

IEC 62037, Preparation of Parts Lists.

IEC 62079, Preparation of Instructions – Structuring, Content and Presentation.

IEC 62424, Representation of Process Control Engineering Requests in P&I Diagrams.

IEC 62491, Labelling of Cables and Cores.

IEC 81346–1, Industrial Systems, Installations and Equipment and Industrial Products – Structuring Principles and Reference Designations – Part 1. Basic Rules.

IEEE Std. 789, Standard Performance Requirements for Communication in High Voltage Environments.
Nibbio, N., Genier, M., Brunner, C., et al. (2010) Engineering approach for the end user in IEC 61850 applications, paper D2/B5–115, *CIGRE session*.
Pereire, A.T.A., Ferreira, J.C. and Tavares, T.N. (December 2012) Lessons learned from the use of IEC 61850 in substation automation systems, *PAC World Magazine*.
Perez, F., Calvo, I., Lopez, F. and Etxeberria-Agiriano, I. (n.d.) Dealing with communication channels within IEC 61499 component-based systems.
Sidhu, T.S., Kanabar, M.G. and Parikh, P.P. (December 2008) Implementation issues with IEC 61850 based substation automation systems, *15th National Power Systems Conference (NPSC)*, IIT Bombay.
UL Std. 1773, Termination Boxes.

10

Communication with the Remote Control Center

Like other geographically dispersed systems, power systems are equipped with SCADA systems (Supervisory Control and Data Acquisition Systems), which consist of a number of electronic devices (slave devices) located at different substations collecting data and sending that data back to a remote master station via a communication system. The master station displays the collected data and also allows the power system operator to perform remote control on substation switchgears.

10.1 Communication Pathway

Due to nature of electrical power systems, communication links between HV substations and the remote control center are active on a permanent basis. Commonly, communication links for long distances are established by using communication media such as telephone line, fiber-optic network, radio and/or microwave transmission resulting in the overall topology shown in Figure 10.1.

10.2 Brief on Digital Communication

Unlike analog communication, in which continuous waves are transmitted, digital communication consists of the transfer of a stream of binary digits (0s and 1s) from one place to another. Because of this, the digital communication process is a more complex process and deals with a significant quantity of issues, such as the following:

- Transmission media
- Mechanism for generation of binary data

Substation Automation Systems: Design and Implementation, First Edition. Evelio Padilla.

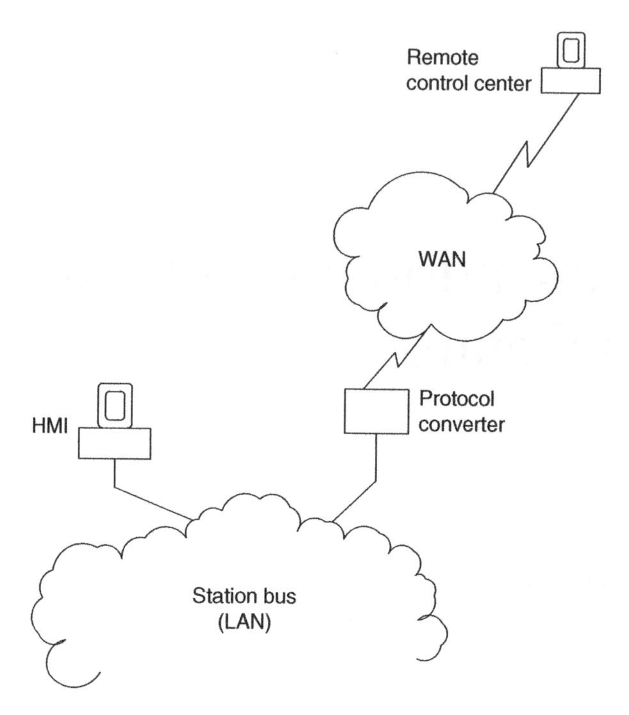

Figure 10.1 Communication pathway with the remote control center

- Sending procedures
- Addressing of target devices
- Coding/decoding techniques
- Error detection techniques
- Encryption methods.

At the beginning of digital age, device manufacturers faced up to different issues when applying their proprietary technologies bringing as a consequence a proliferation of many closed network systems in which only devices coming from a same manufacturer were able to communicate; that was until 1978 when the ISO (International Organization for Standardization) put order to the matter by publishing Standard 7498 in which the model known as the Open Systems Interconnection Reference Model is defined, nowadays known as the seven-layers model.

10.2.1 The OSI Reference Model

The reference model mentioned earlier is based on slicing the communication process up into seven parts called "layers" that pile up in a vertical direction to conform a data transmission management structure configured as shown in Figure 10.2.

These layers represent the resources in terms of functionalities and services that communication devices exhibit to collaborate in order to get efficient communication. Each layer must

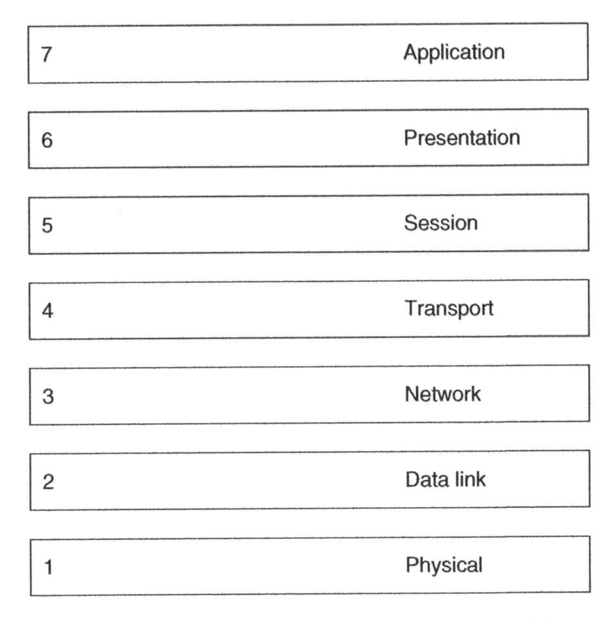

Figure 10.2 Structure of the OSI Reference Model

provide the specific services needed to make data communication successful. The services provided by each layer are performed by "entities" conceivable as abstract devices, like programs and functions that implement the particular service. The means to offer such services are based on hardware and/or software depending on the position of the layer, in such a way that lower layers do their job through hardware or firmware (software that runs on specific piece of hardware) while higher position layers do their work using software.

The reason for the vertical piling is to indicate that each layer interfaces with another for exchanging capabilities and support in such manner that services available on higher layers are the summation of all services provided by the lower layers.

The model brings several significant benefits, which include the following:

- Approach network users with a better understanding of the complex networking technology process.
- Evidence the strength of the conjunction hardware + software to perform specific functions.
- Establish a specialized terminology useful to analyze network functionalities.
- Allow vendors to declare their device networking capabilities in a standardized manner.

The elements and purposes of each layer are the following:

Layer 1 (The Physical Layer) The physical layer is a *hardware interface* with communication media, with respect to physical, electrical and functional definitions. Network components subject to this layer include:

- Serial interfaces
- Coupling modules

- Adapters/terminal accessories
- Cards for network accessing.

Layer 2 (The Data Link Layer) The data link layer deals with the task of *sending packages of data* (structured streams of bits) from one place to another ensuring an error free message at the destination device. Networking devices that operate at the data link layer include:

- Switches
- Bridges
- Modems.

Layer 3 (The Network Layer) The network layer is responsible for *traffic control* into the network, based mainly on the application of a logical addressing system further to other artifices such as the fragmentation of large data packages. The most common networking device that operates at the network layer for communicating separate networks is the router.

Layer 4 (The Transport Layer) The transport layer offers the service of the overall *management of communication* between two specific devices, independent of possible constraints imposed by layers 1–3. This is made through the execution of various tasks like the following:

- Identification of interacting devices.
- Confirmation of the message integrity.
- Segmentation of large data packages.
- Data traffic control.

The most common networking resource that works at the transport layer is the Transmission Control Protocol (TCP).

Layer 5 (The Session Layer) The session layer provides the set of services needed to *establish a dialog* between networked devices. The associated tasks to this layer include:

- Relationship between application programs.
- Data flow control.
- Driving of dialog means.
- Tracking of transferred data.

Layer 6 (The Presentation Layer) The presentation layer deals with *treatment of messages* for security or for transferring efficiency, such as message encryption or message compression.

Layer 7 (The Application Layer) The application layer is responsible for *providing interactive interfaces* with human operators of networked devices.
Examples of functions carried out in this layer are:

- Printing tasks
- Transferences of integrated files
- E-mail services.

User application level
Application layer
Pseudo transport layer
Link layer
Physical layer

Figure 10.3 Structure of the EPA model

10.2.2 The IEC Enhanced Performance Architecture Model

This model, also called the EPA model, was defined by the IEC for telecontrol schemes based on serial communication addressed to fulfill control and monitoring needs of geographically dispersed systems. The model consists of the layers shown in Figure 10.3.

The sets of layers are as follows:

Physical Layer: The physical layer deals with *interfaces* between networked devices and communication media.

Link Layer: The link layer defines *transmission procedures* including transmission modes and frame formats (data packet formats).

Pseudo Transport Layer: The pseudo transport layer is responsible for *message fragmentation* when messages exceed the size of standard data frame.

Application Layer: The application layer establishes the *structure and contents of data fields* of the standard data packets.

User Application Level: The user application level defines the *functions that are performed* by the remote master station.

10.3 Overview of the Distributed Network Protocol (DNP3)

DNP3 is an open protocol available for use by any user or device vendor, developed specifically for SCADA applications that require transmission of small data packets but with a significant reliability level. This protocol, based on the IEC EPA model architecture, defines communications between a central master station and a lot of slave devices located at different substations (formally called outstations). One of the responsibilities of the master station is to keep its database updated in such way that it always reflects the actual status and conditions of substation components.

As a basic operative principle, the master station sends data frames (structured packets of data called *requests*) to slave devices that return other data frames back to the master station (called responses to requests). In addition, slave devices may also send master stations spontaneous frames called *unsolicited responses*.

The specification of the protocol is officially controlled by the DNP Users Group through their website (www.dnp.org) in which several related documents appear including an exhaustive template (DNP3 Specification – Device Profile) to be used by device vendors for declaring the DNP capabilities of their products.

10.3.1 The Device Profile Document

The DNP Users Group offers a Device Profile Template on its website as a standard format for describing the capabilities of any device declared by the manufacturer as DNP3 compliant. The pre-defined information contained in the template to be completed by the vendor includes:

- Identification of the device and subassemblies
- DNP level supported
- Details on connections by serial link and by IP networking
- Details on implementation to different layers
- Features that apply to the master station
- Features that apply to slave devices (outstations)
- Performance of slave devices
- Details about security
- Information related to databases.

The completed Device Profile Document must be provided by all device vendors on user request. When any item is marked up to be configurable, the particular methods to do that must be indicated in the document.

10.3.2 The DNP3 Implementation Level

Looking at the Device Profile Document template provided by the DNP Users Group, the reader may appreciate the tremendous amount of features that are required to be DNP device compliant. However, considering that the full implementation of DNP features has a strong impact on memory requirement and speed of the device, it is recognized that not all devices need the complete set of DNP features. Because of this, the template includes a specific item in which the device vendor must declare a specific level (1, 2, 3 or 4) to which the device conforms. Based on that, typical supported levels finding on the marketplace according to device purpose are.

Level 1: Sensors, transducers and simple meters
Level 2: Intelligent Electronic Devices (IEDs) such as relays and advanced meters, also including devices for use in small-SCADA systems.
Level 3 and 4: Data concentrators and units for large SCADA systems.

Although the levels mentioned represent a general tendency, at least level 3 may be required in a substation slave device when communication with the remote control center includes the transmission of time-stamped events.

10.3.3 The DNP3 Implementation Document

This document provided by the SAS integrator describes the implementation of the DNP3 protocol in a substation slave device (station computer or gateway solution). It has the following objectives:

- Serve as a guideline to the protocol implementation project team.
- Serve as reference for preparation of protocol testing instructions.
- Allow the SAS user to validate protocol features and restrictions, and check during the engineering stage that the protocol implementation fulfills the specified requirements.
- Remain part of "as-built documents" after updates following site testing.

The content of the document includes the following items:

- *Details of implementation*: Any restriction that shall be respected, the Device Profile Document refers to the specific slave device and the Implementation Table.
- *Configuration settings*: Settings for general behavior of the device, these are settings for; analog inputs, binary inputs, analog outputs and binary outputs.
- *Operational aspects*: Startup sequence, modes of interaction with the master station, procedures for time synchronization and the mechanism for handle opening/closing commands by the master station
- *Diagnostic information*: This refers to the means to allow the SAS user to follow the activities that are going on between the slave device and the master station.

Further Reading

Clarke, G. and Wright, E. (2009) *Practical DNP3 and Modern SCADA Systems*, IDC Technologies.

DNP Users Group (2005) A DNP3 Protocol Primer.

DNP3 Specification (Device Profile Document), www.dnp.org.

EPRI (2012) Distributed Network Protocol (DNP3) Security Interoperability Testing 2012, report 1026561.

IEC 60870-5-101, Telecontrol Equipment and Systems – Part 5-101 Transmission protocols – Companion Standard for Basic Telecontrol Tasks.

IEC 60870-5-SER, Telecontrol Equipment and Systems – Part 5: Transmission Protocols – ALL PARTS.

IEEE 1815, Standard for Electric Power Systems Communications – Distributed Network Protocol (DNP3).

ISO/IEC 7498-1, Information Technology – Open Systems Interconnection – Base Reference Model: The Basic Model – Part 1.

Lebow, I. (1998) *Understanding Digital Transmission and Recording*, IEEE Press, Piscataway.

Makhija, J. (November 2003) Comparison of protocols used in remote monitoring: DNP 3.0, IEC 870-5-101 & Modbus, *M. Tech, Credit Seminar, Electronics Systems Group*, EE Dept, IIT Bombay.

11

System Attributes

Each particular Substation Automation System (SAS) will have a lot of attributes imposed by different factors such as user requirements, provider philosophy, hardware/software capabilities and the state-of-the-art in technology when they are implemented. This chapter gives an overview of several of them intending to provide guidelines for engineers dealing with SASs in different project stages.

11.1 System Concept

Modern SASs are designed using a decentralized automation concept based on a hierarchical structure conformed by the following control levels:

- Remote control level
- Station level
- Bay level
- Process level.

Such a concept is characterized by the intensive use of Intelligent Electronic Devices (IEDs) and the implementation of optical fiber links for data communication between IEDs. The main benefits that are offered by the current SAS generation include:

- Elimination of electromagnetic inductions by using optical transmission techniques for control related signals.
- Elimination of intermediary and supplementary elements such as transducers and sequence of events recorders.
- Reduction of overloads on communication channels between bay control level and station control level.

Substation Automation Systems: Design and Implementation, First Edition. Evelio Padilla.

- Reduction of hardwire cabling.
- Simple and easy installation of control devices and related assemblies.
- Flexible design of control functions based mainly in software.
- Permanent monitoring through the use of software instead of separate supervisory means.
- Minimal maintenance needs.
- Permanent monitoring of the state of HV and MV switchgear and power transformers.
- Effective presentation of arrangement and status of primary circuit and equipment.

The control network itself is powered by the integration of compatible protective relays, which are also able to effectively deliver current and voltage values in an appropriate form for transfer to a higher hierarchical level, as well communicate horizontally with bay controllers. As an additional advantage, the measurement functions are performed by bay controllers. In the favor of reliability, a watchdog capability and other similar continuous monitoring functions are implemented in the bay controllers.

The bay controllers are built in a modular manner (typically 48 cm/19" in size) and in some cases use components readily available on the market. The fiber-optic links may be designed for use with standard easy to install connectors. The software packets are designed on a modular basis and use higher level programming languages.

11.2 Network Topology

The main communication network of the SAS (station bus) connects all of the IEDs to one another (bay controllers, protective relays, station controller, etc.) and to a modem or other devices for communication outside the substation. Such a network may be configured in different ways considering several factors like reliability, security, speed and maintainability. Because the intent to simultaneously optimize all these mentioned factors may create conflict in the network design, in a theoretical sense, network designers try to create isolated network segments to get security, redundant schemes for enhanced availability and single network segments to obtain high speed communication. Although in practice SAS providers have their own typical arrangements, they try to adapt to the particular needs of each specific SAS project. The options to configure the station bus include those in Figures 11.1–11.4.

In order to give guidelines on strengths and weaknesses of different network arrangements, a summary is shown in Table 11.1.

Appropriate network topology should be defined for each specific SAS project based on the required data handling, quality requirements according to the operative significance of the substation, commissioning requirements during the installation of the SAS and cost impact. Essential factors to consider in the process of topology definition include:

- Make sure that distributed functions are fully defined and have acceptable effects in cases of degraded network condition.
- Manage the increasing system complexity when additional components are added.
- Keep in mind key features like interoperability and system performance.
- Be sure you have the complete overview of the SAS components, including collateral components such as equipment belonging to the auxiliary power system, separate disturbance recording and an external alarm annunciator.

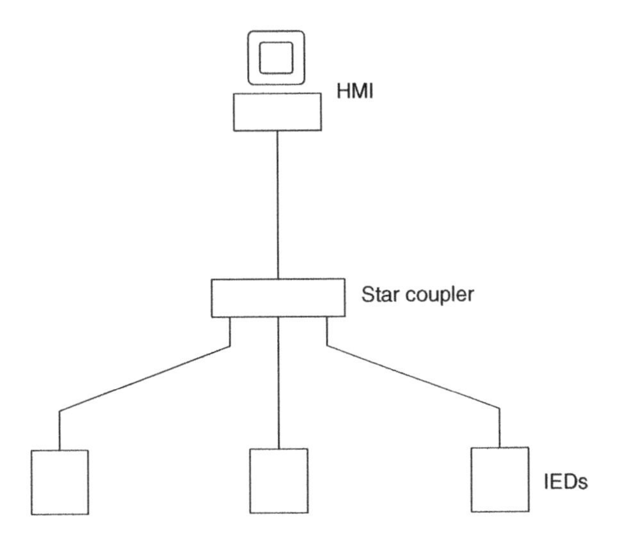

Figure 11.1 Single star topology

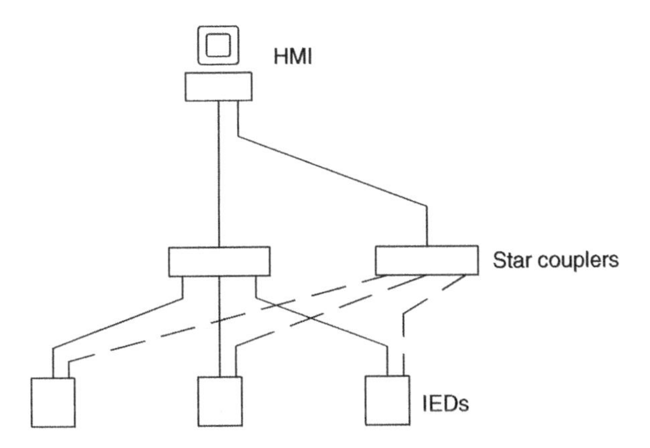

Figure 11.2 Double star topology

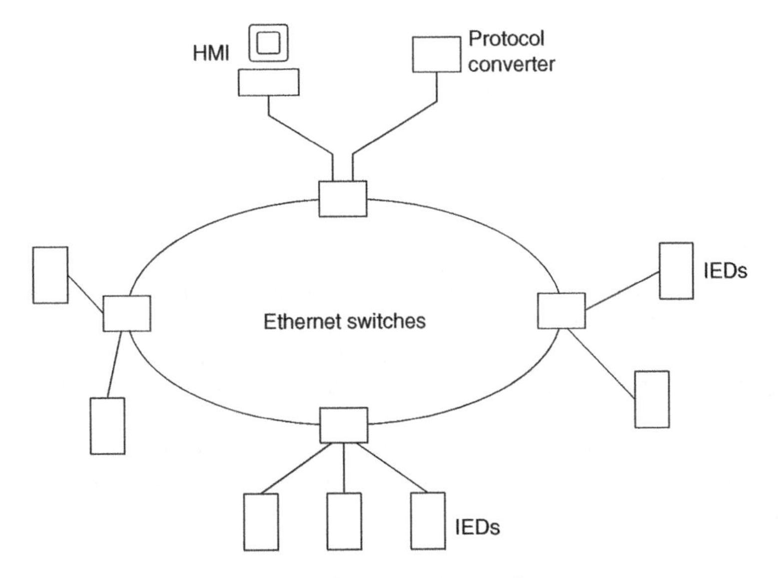

Figure 11.3 Single ring topology

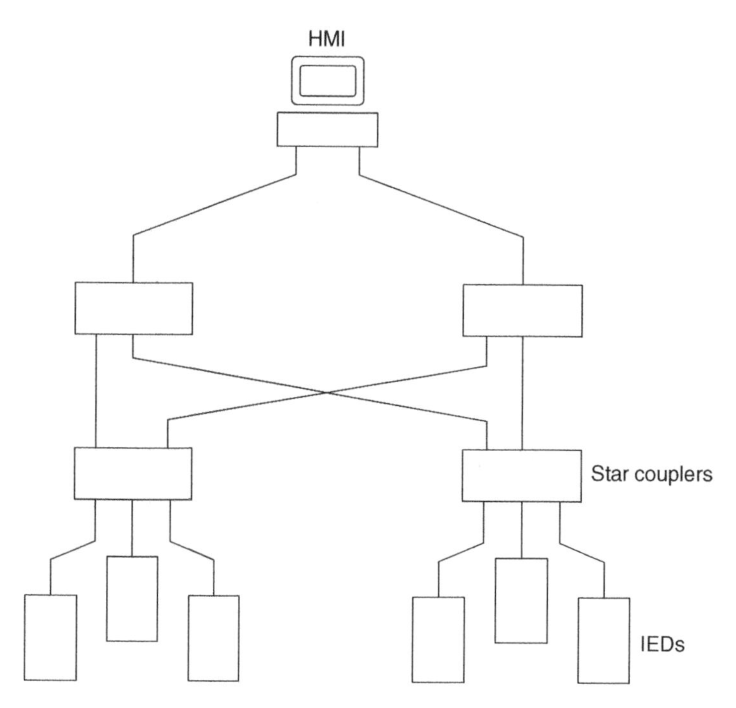

Figure 11.4 Crossover topology

Table 11.1 Comparison between several network arrangements

Intrinsic feature	Network topology			
	Single star	Double star	Single ring	Crossover
Redundancy degree	low	high	high	medium
Performance in data transferring process	high	high	medium	high
Diagnostic feasibility	high	high	medium	high
Flexibility for expandability	high	high	medium	high
Flexibility for connectivity	high	high	medium	high
Performance in transferring time-critical messages	high	high	low	high

11.3 Redundancy Options

People like redundancy, but it means extra cost as well as more complexity in implementation. Redundant arrangements involve establishing duplicate elements in specific positions of a SAS network to ensure that a service is still offered even when one of those elements is faulty. Redundancy may be implemented in different segments of the SAS such as follows:

- *Redundancy in the bay controller*: The complete set for local control (bay controller more related accessories) can be doubled in a hot/standby configuration when a high availability is required, for example in EHV substations.

- *Redundancy in device connectivity*: This refers to duplicate devices that act as common node in the network, such as star coupler devices.
- *Redundancy in station controller/HMI*: In a SAS where extremely high availability is required, the concept of a hot/standby base system may be applied. This concept is based on data shadowing of disk-resident data as well as RAM-resident data between two base systems. The shadowing task is made by the application software installed into the station controller. In case of a failure of the base system when in a hot condition, a switchover takes place meaning that the unit receiving the shadowed data in standby mode becomes the new active unit.
- *Redundancy in communication media*: The purpose of redundancy in communication media is to prevent a single point of failure. In case of failure in a main link, the network must recover within a given short time by re-routing data traffic through the second link. When this type of redundancy is applied, a redundancy control protocol is needed to manage the communication links.
- *Redundancy in power supplies*: This redundancy is recommended for all critical devices like bay controllers, star coupler devices and station controllers.
- *Redundancy in other internal components*: For extremely critical devices, such as star coupler devices, further to redundancy in power supplies, it is reasonable to also ask for redundancy in other important elements, for example the bus-administrator module.
- *Redundancy in power circuits*: It is essential that the auxiliary power system of the substation can provide separate DC power circuits to serve redundant power sources of different devices.

In addition to the mentioned redundancy options, care must be taken with respect to connecting main and backup IEDs to connectivity devices, such as star couplers, through separate (independent) connection modules.

11.4 Quality Attributes

The SAS plays a critical role in the overall mission for efficient operation of the power system in a secure manner, as well as to maintain the electrical service provided to loads connected to the substation. Because of this, the SAS must be robust enough to satisfy expected performance with minimum effort from the substation owner in terms of intervention, maintenance and repair activities. The way to guarantee such capability consists of analysis during the engineering stage so that the proposed SAS exhibits various quality attributes, which include those in the following subsections.

11.4.1 System Reliability and Availability

Many factors affect the reliability and availability of a SAS. Usually, redundancy is the first thing to reflect upon, but it is never the most important factor. The basic system proprieties are normally much more important. Those system proprieties come from the following aspects:

- Robust mechanical and electrical design of the components.
- Immunity against electromagnetic interference.

- High quality components and connections.
- Modularized and well established hardware.
- Modular and well tested software.
- Comprehensible programming language.
- Detailed engineering documentation.
- Built-in self-supervision facilities.
- Appropriate security mechanisms.
- Adequate knowledge level of SAS provider in substation process.
- Good understanding of the distributed control concept by the SAS provider.

11.4.1.1 Considerations of the Standards

Current standards (IEC 61850) allow SAS reliability performance according to the "graceful degradation" principle, which means the ability of the system to maintain limited functionality even when a significant portion of it is faulty. In such conditions some operative parameters of the system (e.g., transmission speed) will decline gradually as an increasing number of components become faulty. Concerning availability, reference is made to the Standard IEC 60870–4, which establish three availability classes: A1 (99.00%), A2 (99.75%) and A3 (99.95%).

11.4.1.2 Example of an Availability Calculation

As a guideline to predict availability of a proposed system, an example based on standards MIL-STD-785 and MIL-HDBK-217 follows:

Relevant definitions:

MTTF(h): Mean Time To Failure (in hours)

MTTR(h): Mean Time To Repair considering all components of the system (in hours)

A: Availability

Applicable formulas:

MTTF(Seg.) = MTTF1 + MTTF2 (if the components in the segment are connected in redundant arrangement)

1/MTTF (Seg.) = 1/MTTF1 + 1/MTTF2 (if the components in the segment are connected in series)

A = MTTF/(MTTF + MTTR)

Topology to be analyzed:

The analyzed topology is the hypothetical one shown in Figure 11.5.

Component reliability figures:

Device vendors must provide specific MTTF related their products as well as an indication of the method used for calculation. Fictitious values are shown in Table 11.2 purely for didactic purposes.

Premises:

An assumption of MTTR = 7 hours is made including trouble locating and component replacement.

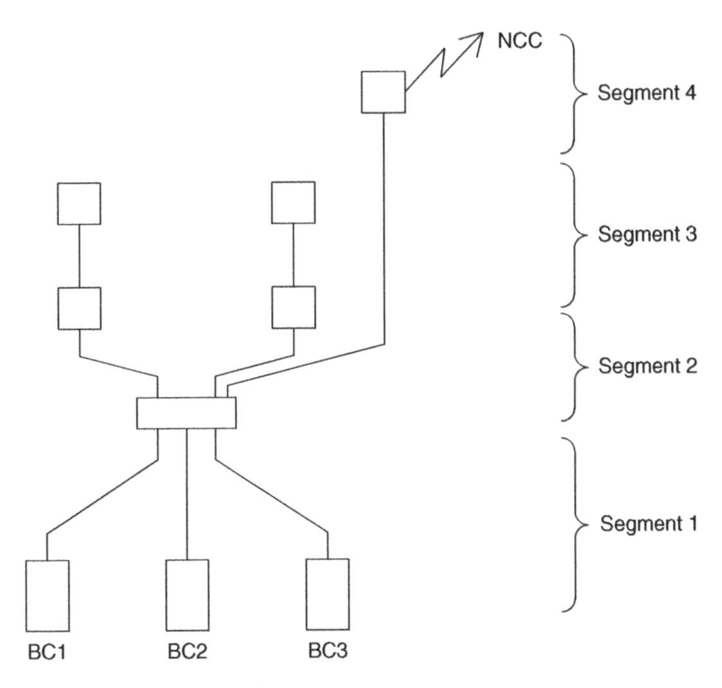

Figure 11.5 Hypothetical SAS topology for availability calculation

Table 11.2 Summary of component reliability figures

Component	Details	MTTF(h)
Bay controller	With redundant power supply	950,000
Communication link	Optical media	12,000,000
Star coupler	With redundant components	1,000,000
Station controller	Industrial grade computer	200,000
HMI (Human Machine Interface)	Including monitor, keyboard and mouse	100,000
Protocol converter	With redundant CPU	300,000

Segment reliability models:

Each segment is previously evaluated as follows:

Reliability model for segment 1 (related to bay controller)

Bay controller MTTF(h): 950,000	Communication link MTTF(h): 12,000,000

MTTF seg. 1 (h) = 880,000

Reliability model for segment 2 (related to star coupler)

Star Coupler MTTF(h): 1,000,000	Communication link MTTF(h): 12,000,000

MTTF seg. 2 (h) = 923,000

Reliability model for segment 3 (related to station controller/HMI)

Station Controller 1 MTTF(h): 200,000	Communication link 1 MTTF(h): 12,000,000	HMI 1 MTTF(h): 100,000
Station Controller 2 MTTF(h): 200,000	Communication link 2 MTTF(h): 12,000,000	HMI 2 MTTF(h): 100,000

MTTF seg. 3 (h) = 132,596

Reliability model for segment 4 (related to protocol converter)

Protocol converter MTTF(h): 300,000	Communication link MTTF(h): 12,000,000

MTTF seg. 4 (h) = 292,688

Overall system reliability model:

Segment 1 MTTF(h): 880,000	Segment 2 MTTF(h): 923,000	Segment 3 MTTF(h): 132,596

Segment 4 MTTF(h): 292,688

This is equivalent to:

Segment 1 MTTF(h): 880,000	Segment 2 MTTF(h): 923,000	Segment 3 + Segment 4 MTTF(h): 427,284

MTTF overall system (h) = 219,298
MTTR(h) = 6

Results:
A = 219,298/(219,298 + 7)
A = 0.9999
A = 99.99%

This result satisfies the minimum acceptable 99.95% for standardized availability class A3.

11.4.2 System Maintainability and Security

Maintainability means the probability of repairing the system during a given time, while security refers to the probability of sending spurious control commands. Both attributes are governed by the Standard IEC 60870–4.

11.5 Provisions for Extendibility in Future

If the substation is going to be extended in future, the SAS must be suitable for extension by adding new bays/feeders. That means that all drawings and configurations, alarm/event list and other relevant facilities shall be designed in such way that its extension shall be easily performed by the user. The provisions include that normal operation of the existing substation must be unaffected and the system must not require a shutdown. In such cases, the contract scope will cover all necessary software tools along with source codes to perform addition of bays/feeders in future and complete integration with existing portions of the SAS. The software tools must be able to configure IEDs, add additional analog variables, modify alarm/event lists and change/add interlocking conditions relating to the extension.

11.6 Cyber-Security Considerations

Cyber-security is an issue of increasing concern for utilities and SAS providers. This is because, with time, cyber-attacks on public utilities, including the energy sector, are becoming more and more frequent. Vulnerability come from the fact that SASs are interconnected with remote control centers by wide-area communication networks, further, most of them are equipped with a communication port that allows connection to the substation via the Internet with a remote engineering station belonging to the SAS provider (for technical service). There are also utilities that implement Internet portals that enable workers to access their control systems. Threats may come from sophisticated external hacking groups but also from disgruntled employees or from unauthorized persons who know the weaknesses in the internal operating rules and procedures.

To achieve cyber-security protection in terms of confidentiality, integrity and availability of the electronic information communication system, contributions of different SAS project players are essential. The substation owner may contribute by defining a comprehensive set of cyber-security requirements early on, for example the following:

- Asking for the proposed system topologies the cyber-security factor is being considered by, for example, establishing multiple secure zones.
- Requesting appropriate protection for the electronic security perimeter.
- Including robustness and hardening features for the communication network.
- Demanding the implementation of authorization and authentication mechanisms.
- Requesting auditability and logging facilities
- Asking for antivirus protection.

Device vendors contribute by developing their products following well-structured processes, which include cyber-threat modeling and fixing security features in devices like proper access media and security logging. Finally, the system integrator contributes by ensuring that all the cyber-security capabilities of system devices are used and configured in an effective manner on the overall system.

11.7 SAS Performance Requirements

There are a lot of specific requirements that may be established by substation owner in their specification. These requirements include the following:

- *Time over for control command*: This refers to the requirement that if a control action is not completed within a specified time, the command should be cancelled (e.g., 5 s).

- *Remaining memory capability for event storage in IEDs*: When the memory is loaded at a certain percentage, the system sounds an alarm (e.g., 75%).
- *Updating analog values shown on HMI*: This must be made at a regular predefined time (e.g., every 2 ms).
- *Time stamping resolution for events and alarms*: Resolution must have a maximum time limit (e.g., 1 ms).
- *Capability for event recording by IED*: This must be enough to store all events occurring during a certain time period (e.g., at least 1 month).
- *Maximum load in CPU and RAM of station controller*: These internal components are selected in such manner that during normal operation no more than a certain percentage in capacity of processing and memory is used (e.g., 40%).
- *Storage capability of bay controllers*: This is the expectation of bay controllers to have capability to store all the managed data for the minimum of a pre-defined time (e.g., 48 hours).

Further Reading

Anderson, D. and Kipp, N. (2010) *Implementing Firewalls for Modern Substation Cybersecurity*, PAC World Conference.

Andersson, L., Brand, K-P., Brunner, C. and Wimmer, W. (June 27–30, 2005) Reliability investigations for SA communication architectures based on IEC 61850, paper 604 in the Poster Session at IEEE St. *Petersburg PowerTech*, St. Petersburg.

Benlamkaddem, F., Faija, A., Grillo, P., et al. (2012) Cyber Security: System Services for the Safeguard of Digital Substation Automation Systems, paper D2–102, *CIGRE session.*

Botza, Y., Shaw, M., Allen, P., Staunton, M. and Cox, R., *Configuration and Performance of IEC 61850 for First-Time Users – UNC Charlotte Senior Design Project*, paper by University of North Carolina and Schweitzer Engineering Laboratories, Inc.

Brand, K-P., Riemann, P., Maeda, T. and Wimmer, W. (2006) Requirements of interoperable distributed functions and architectures in IEC 61850-based SA systems, paper B5–110, *CIGRE session*.

Buecker, A., Andreas, P. and Paisley, S. (2008) Understanding IT perimeter security, paper for IBM.

Cyber Security Considerations in Power System Operations, *Electra Review* 218, 15–22, February 2005.

Galea, M. and Pozzuoli, M., *Redundance in Substation LANs with the Rapid Spanning Tree Protocol* (IEEE 802.1w), paper for RuggedCom Inc.

Gauci, A., Garratano, D. and Pathania, S. (2014) *A Framework for Developing and Evaluating Utility Substation Cyber Security*, report for Schneider Electric.

Hancock, W. (1988) *Designing and Implementing Ethernet Networks*, John Wiley & Sons, Ltd, Chichester.

Hohlbaum, F., Schwyter, P. and Alvarez, F. (October 19–20, 2011) Cyber Security requirements and related standards for Substation Automation Systems, *CIGRE Colloquium*, Buenos Aires.

IEC 62351-SER, Power System Management and Associated Information Exchange – Data and Communication Security – ALL PARTS.

IEEE 1686, Standard for Intelligent Electronic Devices, Cyber Security Capabilities.

MIL-HDBK-217, Reliability Prediction of Electronic Equipment.

MIL-STD-785, Reliability program for Systems and Equipment Development and Production.

Sidhhu, T.S., Kanabar, M.G. and Parikh, P.P., (2008) implementation issues with IEC 61850 Based Substation Automation Systems, *Fifteenth National Power Systems Conference (NPSC)*, IIT Bombay, December 2008.

Steinhauser, F., Schossig, T., Klin, A. and Geiger, S. (2010) Performance Measurements for IEC 61850 and Systems, PAC World Conference.

UCA International Users Group (2012) Security Profile For Substation Automation.

Vadiati, M., Shariati, M.R., Farzalizadeh, S., et al. (2010) Buses architecture of substation automation system based on significance level of substation, *Journal of Applied Sciences* 10, 2464–2468.

Wester, C. and Adamiak, M., Practical Applications of Ethernet in Substations and Industrial Facilities, Paper for GE Digital Energy Multilin.

12

Tests on SAS Components

Before electrical components can be installed in an electrical system in a safe and reliable manner, they must pass a number of tests to ensure that intended service will be successful fulfilled during the complete life-cycle. Being complex artifacts, SAS devices are subjected to extensive testing programs, far more than other substation components. The testing programs include:

- Type tests:
  - Basic characteristics tests
  - Functional tests
- Acceptance tests
- Tests for checking the compliance with Standard IEC 61850.

Scopes and details of such tests follow.

12.1 Type Tests

Type tests are those tests that are performed once on a type of product to establish that the product meets applicable industry standards. The tests are performed by either the product manufacturer or by an independent test authority.

A type test program on IEDs includes verification of basic characteristics and also functional tests, such as described later for the bay controller as a representative piece of essential SAS devices.

12.1.1 Basic Characteristics Tests

Basic characteristics tests allow checking that a tested device exhibits the required strengths to operate satisfactorily in the environment in which it will be exposed. The tests applicable to devices that will be dedicated as bay controllers are listed in Table 12.1.

Substation Automation Systems: Design and Implementation, First Edition. Evelio Padilla.

Table 12.1 Basic characteristics tests for bay controllers

Climatic Environment		
Test	**Standard**	**Details**
Temperature		
• Test A: Cold	IEC 60068-2-1	−10°C/1 d
• Test B: Dry heat	IEC 60068-2-2	+55°C/1 d
• Test N: Change of temperature	IEC 60068-2-14	−10°C–+55°C
Humidity	IEC 60068-2-30	92%, variant 1, 4 d
Mechanical Environment		
Test	**Standard**	**Details**
Vibration		
• Response	IEC 60255–21–1	Up to 150 Hz and 0.5 gn
• Endurance	IEC 60255–21–1	Up to 100 Hz and 1.0 gn
Shock and bump		
• Response	IEC 60255–21–2	18 × 5 gn, 10 ms
• Withstand	IEC 60255–21–2	18 × 15 gn, 10 ms
• Bump	IEC 60255–21–2	1000 × 10 gn, 15 ms
Seismic	IEC 60255–21–3	Up to 35 Hz, 1.0 gn vertical Up to 35 Hz, 2.0 gn horizontal
Electrical Environment		
Test	**Standard**	**Details**
Voltage interruptions	IEC 60255–26	40 ms
Voltage ripple on DC	IEC 60255–26	12%
Clearance and creepage distance	IEC 60255–27	
Insulation resistance	IEC 60255–27	500 V, more than 100 MOhm
Dielectric withstand at power frequency	IEC 60255–27	2kV, 1 min
Impulse voltage	IEC 60255–27	5 kV, 1.2 × 50 uS
Insulation across open relay contacts	IEC 60255–27	1 kV AC, 1 min
Burst disturbance immunity	IEC 61000–4-4	1 MHz, 400 Hz
Electrostatic discharge immunity	IEC 61000–4-2	8 kV air, 6 kV contact
Electrical fast transient	IEC 61000–4-4	4 kV, 5 kHz
Surge immunity	IEC 61000–4-5	1.2 × 50 uS, 2 kV 8 × 20 uS, 4 kV
Radiated radio-frequency, electromagnetic field immunity	IEC 61000–4-3	10 V/m, 80–2000 MHz
Power frequency magnetic field immunity	IEC 61000–4-8	
Immunity to conducted disturbances, induced by radio-frequency	IEC 61000–4-6	
Conducted and radiated RF emission	IEC 60255–26	30–230 MHz 230–1000 MHz
Current transformers		
• Thermal rating continuous	IEC 60255–1	30 × In 10 s
• Short duration overload		4 × In continuous
• Power consumption at In		

Table 12.1 (*continued*)

Climatic Environment		
Test	Standard	Details
Voltage transformers		
• Thermal rating continuous	IEC 60255-1	2 × Vn continuous
• Short duration overload		3 × Vn 10 s
• Power consumption at Vn		
Contact rating		
• Maximum operating voltage		
• Continuous rating		
• Surge withstand		
• DC making capacity		
• Making power		
• DC breaking capacity		
Power consumption		At Vn DC
• In normal condition		
• In operating condition		

Table 12.2 Functional tests for bay controllers

Test	Standard/procedure	Details
Limited DC ramp	Check of supply voltage rising performance	From 0 to 100% Ramp duration: 1 min
Limited DC power supply interruption	Check of behavior under short time voltage interruption	Minimum allowed interruption time: 50 ms
AC Component (Ripple) on DC supply voltage		Inject at least a ripple amplitude of 12%
Thermal loadability of current transformers		4 × In, 12 h
Thermal loadability of voltage transformers		2 × Vn, 12 h
Measuring comparison between injected and indicated (active power)	IEC 60688	Maximum allowed difference: 1.0%
Measuring comparison between injected and indicated (reactive power)	IEC 60688	Maximum allowed difference: 2.0%

12.1.2 Functional Tests

Functional tests have the purpose of verifying that a tested device will show the expected behavior under relevant operating conditions during its real performance. Typical functional tests carried out on IEDs used as bay controllers are indicated in Table 12.2.

12.2 Acceptance Tests

Acceptance tests (also called routine tests or production tests) are performed on each unit of a type of product to ensure a constant quality of delivered units in accordance with the quality system of the manufacturer. The scope typically covers dielectric tests and visual checking tests.

12.3 Tests for Checking the Compliance with the Standard IEC 61850

The framework to verify the compliance of a device with the standard IEC 61850 is given in part 10 of the standard (IEC 61850–10: Conformance Testing).

The purpose of that testing program is to determine whether a device conforms to the definitions of the standard. This is made by exchange messages between a test system and the device under test. It is also expected that standard compliance improves the chance to get device interoperability.

Further Reading

Brunner, C. and Apostolov, A. (December 2010) Functional Testing of IEC 61850 based systems, *PAC World* magazine.

CIGRE (October 2003) Conformance Testing Guideline for Communication in Substation, *Technical Report* 236.

IEC 60068-2-1, Environmental Testing – Part 2–1: Test A: Cold

IEC 60068-2-14, Environmental Testing – Part 2–14: Test N: Change of Temperature.

IEC 60068-2-2, Environmental Testing – Part 2–2: Test B: Dry Heat

IEC 60255–1, Measuring Relays and Protection Equipment – Part 1: Common Requirements.

IEC 60255–21–1, Measuring Relays and Protection Equipment – Part 21–1: Vibration, Shock, Bump and Seismic Tests on Measuring Relays and Protection Equipment – Vibration Tests (Sinusoidal).

IEC 60255–21–2, Measuring Relays and Protection Equipment – Part 21–1: Vibration, Shock, Bump and Seismic Tests on Measuring Relays and Protection Equipment – Shock and Bump Tests.

IEC 60255–21–3, Measuring Relays and Protection Equipment – Part 21–1: Vibration, Shock, Bump and Seismic Tests on Measuring Relays and Protection Equipment – Seismic Tests.

IEC 60255–26, Measuring Relays and Protection Equipment – Part 26: Electromagnetic Compatibility Requirements.

IEC 60688, Electrical Measuring Transducers for Converting A.C. and D.C. Electrical Quantities to Analogue or Digital Signals.

IEC 61000–1-2, Electromagnetic Compatibility (EMC) - Part 1–2: General - Methodology for the Achievement of Functional Safety of Electrical and Electronic Systems Including Equipment with Regard to Electromagnetic Phenomena

IEC 61000–4-11, Electromagnetic Compatibility (EMC) – Part 4–11: Testing and Measuring Techniques – Voltage Dips, Short Interruptions and Voltage Variations Immunity Test.

IEC 61000–4-2, Electromagnetic Compatibility (EMC) – Part 4–2: Testing and Measurement Techniques – Electrostatic Discharge Immunity Test.

IEC 61000–4-3, Electromagnetic Compatibility (EMC) – Part 4–3: Testing and Measuring Techniques – Radiated, Radio-Frequency, Electromagnetic Field Immunity Test.

IEC 61000–4-4, Electromagnetic Compatibility (EMC) – Part 4–4: Testing and Measuring Techniques – Electrical Fast Transient/Burst Immunity Test.

IEC 61000–4-5, Electromagnetic Compatibility (EMC) – Part 4–5: Testing and Measuring Techniques – Surge Immunity Test.

IEC 61000–4-6, Electromagnetic Compatibility (EMC) – Part 4–6: Testing and Measuring Techniques – Immunity to Conducted Disturbances, Induced by Radio-Frequency Fields.

IEC 61000–4-8, Electromagnetic Compatibility (EMC) – Part 4–8: Testing and Measuring Techniques – Power Frequency Magnetic Field Immunity Test.

IEC 61850–10, Communication Networks and Systems for Utility Automation – Part 10: Conformance Testing.

IEEE Std. 299, Standard Method for Measuring the Effectiveness of Electromagnetic Shielding Enclosures.

IEEE Std. 82, Standard Test Procedure for Impulse Voltage Tests on Insulated Conductors.

NEMA Std. 250, Enclosures for Electrical Equipment (1000 Volts Maximum).

Udren, E.A. (2006) IEC 61850: Role of Conformance Testing in Successful Integration, Paper for Schweitzer Engineering Laboratories, Inc.

13

Factory Acceptance Tests

The integrating phase of the SAS is concluded by the Factory Acceptance Tests (FATs) to determine that the complete system operates according to the properties agreed in the contract between the supplier and the buyer and covers all functional requirements with acceptable performance. Considering the distributed structure of the system at the switchgear and bay control level, for practical reasons, the tests are conducted on certain portions of the system replacing the remainder with test simulators. It is recommended that the test arrangement contains at least one unit of each and every type of device incorporated in the system to be delivered.

The aim of this chapter is to describe an outline of FAT scope and preparations.

13.1 Test Arrangement

Before starting the FAT it must be ensured that tests will be made on the appropriate system configuration representative of a real situation in definitive system.

13.2 System Simulator

The test simulator that will represent the missing parts of the system must be able to generate all needed inputs for creating certain system conditions, including the heavy load condition into the communication infrastructure.

13.3 Hardware Description

This refers to an available document containing the following information about all components to be part of the test arrangement:

Substation Automation Systems: Design and Implementation, First Edition. Evelio Padilla.

- Component name
- Model
- Serial number
- Manufacturer
- Function/position in the system.

13.4 Software Identification

This is a list that must be prepared in advance related to all software installed in the components that are part of the test arrangement:

- Software name
- Software version
- Device in which the software is installed.

13.5 Test Instruments

A list of instruments and accessories to be used in the tests shall include:

- Instrument name
- Main functions
- Model
- Manufacturer
- Serial number.

13.6 Documentation to be Available

A series of documents must be available at the testing place. These include:

- Test program shown in chronological order of the test's scope.
- Test procedures giving a clear understanding of the tests to be done.
- Test protocol (blank format).
- Technical manuals of all system components that are part of the FAT.
- Software documents.
- Manuals of testing tools.
- Instruction manual for test instruments.
- Applicable technical standards.

13.7 Checking System Features

As a preliminary activity before starting the test program itself, the following checking is recommended:

13.7.1 Checking Basic Features

- System start-up and shutdown
  - Behavior of HMI during start-up process: OK ___ Not OK ___
  - Entrance dialog box appear: OK ___ Not OK ___
  - Behavior of HMI during shutdown process: OK ___ Not OK ___
- Overview of main menu screen
  - Certain in date and time: OK ___ Not OK ___
  - Result of clicking on "Help" key: OK ___ Not OK ___
  - Result of clicking on "Options" key: OK ___ Not OK ___
  - Result of exploring alarm display screen: OK ___ Not OK ___
  - Result of exploring event display screen: OK ___ Not OK ___
  - Result of clicking on other keys: OK ___ Not OK ___.

13.7.2 Checking Power Circuit Screens

- First voltage level (e.g., 230 kV):
  - Appearance of single-line diagram: OK ___ Not OK ___
  - Colors on single-line diagram: OK ___ Not OK ___
  - Effective switchgear status indication: OK ___ Not OK ___
  - Bay level local/remote indication present: OK ___ Not OK ___
  - Switchgear level local/remote indication: OK ___ Not OK ___
  - Measurement indications present: OK ___ Not OK ___
  - Control dialog boxes appear by request: OK ___ Not OK ___
  - Control function procedure: OK ___ Not OK ___
  - Tap changer control dialog box appear: OK ___ Not OK ___
- Second voltage level (e.g., 115 kV):
  - Appearance of single-line diagram: OK ___ Not OK ___
  - Colors on single-line diagram: OK ___ Not OK ___
  - Effective switchgear status indication: OK ___ Not OK ___
  - Bay level local/remote indication present: OK ___ Not OK ___
  - Switchgear level local/remote indication: OK ___ Not OK ___
  - Measurement indications present: OK ___ Not OK ___
  - Control dialog boxes appear by request: OK ___ Not OK ___
  - Control function procedure: OK ___ Not OK ___.
- Auxiliary power system:
  - Appearance of single-line diagram: OK ___ Not OK ___
  - Colors on single-line diagram: OK ___ Not OK ___
  - Effective switchgear status indication: OK ___ Not OK ___
  - Transfer switch manual/auto indication: OK ___ Not OK ___
  - Switchgear local/remote indication: OK ___ Not OK ___
  - Measurement indications present: OK ___ Not OK ___
  - Control dialog boxes appear by request: OK ___ Not OK ___
  - Control function procedure: OK ___ Not OK ___.

13.7.3 Checking the SAS Scheme Screen

- Overall system:
  - Appearance of the set of devices: OK ___ Not OK ___
  - Device identification: OK ___ Not OK ___
  - Interconnecting links: OK ___ Not OK ___
  - Readability: OK ___ Not OK ___
  - Self-supervision functions: OK ___ Not OK ___
- Particular devices (for each one, click on device symbol shown on the overall system screen):
  - Device identification: OK ___ Not OK ___
  - Hardware composition: OK ___ Not OK ___
  - Interconnecting links: OK ___ Not OK ___
  - Software names and versions: OK ___ Not OK ___
  - Parameterization facilities: OK ___ Not OK ___.

13.7.4 Checking Reports Screens (Each Type)

- Active energy report:
  - General appearance: OK ___ Not OK ___
  - Certain in analog values: OK ___ Not OK ___
  - Scales of parameters are adequate: OK ___ Not OK ___
  - Handling tools are enough and efficient: OK ___ Not OK ___
- Reactive energy report:
  - General appearance: OK ___ Not OK ___
  - Certain in analog values: OK ___ Not OK ___
  - Scales of parameters are adequate: OK ___ Not OK ___
  - Handling tools are enough and efficient: OK ___ Not OK ___.

13.7.5 Checking Measurement Screens

- Measurement on first voltage level (e.g., on 230 kV):
  - Structure of values presentation: OK ___ Not OK ___
  - Date and time appear: OK ___ Not OK ___
  - Readability: OK ___ Not OK ___
  - Bay identification appear: OK ___ Not OK ___
  - Provisions for calculate mean values: OK ___ Not OK ___
- Measurement on second voltage level (e.g., on 115 kV):
  - Structure of values presentation: OK ___ Not OK ___
  - Date and time appear: OK ___ Not OK ___
  - Readability: OK ___ Not OK ___
  - Bay identification appear: OK ___ Not OK ___
  - Provisions for calculate mean values: OK ___ Not OK ___
- Measurement on auxiliary power system:
  - Structure of values presentation: OK ___ Not OK ___
  - Date and time appear: OK ___ Not OK ___

  - Readability: OK ___ Not OK ___
  - Feeder identification appear: OK ___ Not OK ___
  - Provisions for calculate mean values: OK ___ Not OK ___.

13.7.6 Checking Time Synchronization Facilities

- Devices and interfaces: OK ___ Not OK ___
- Signal distribution concept: OK ___ Not OK ___.

13.7.7 Checking of Self-Supervision Functions

- Review of all implemented functions: OK ___ Not OK ___
- Texts in messages: OK ___ Not OK ___
- Role of external alarm annunciator: OK ___ Not OK ___
- Means of alarm acknowledgement: OK ___ Not OK ___.

13.7.8 Checking Peripheral Devices

- Printer server: OK ___ Not OK ___
- Hardcopy printer: OK ___ Not OK ___
- Event data logger: OK ___ Not OK ___
- External alarm annunciator: OK ___ Not OK ___.

13.7.9 Checking Collateral Subsystems

- Protective relays:
  - Communication features: OK ___ Not OK ___
  - Relay setting from control devices: OK ___ Not OK ___
  - Event recording when relay act: OK ___ Not OK ___
  - Independent operation: OK ___ Not OK ___
- Disturbance recorder:
  - Communication features: OK ___ Not OK ___
  - Signals that activate the recorder: OK ___ Not OK ___.

13.7.10 Checking Redundant Functionalities

- Redundancy at bay level:
  - Events that motivate switchover OK ___ Not OK ___
  - Prove that no event losses occur OK ___ Not OK ___
- Redundancy at HMI level:
  - Events which motivate switchover OK ___ Not OK ___
  - Prove shadowing of data bases OK ___ Not OK ___.

13.8 Planned Testing Program for FAT

The recommended testing program is presented in the following sections.

13.8.1 System Behavior in an Avalanche Condition

- All expected events appear: OK ___ Not OK ___
- All expected alarms appear: OK ___ Not OK ___.

13.8.2 System Performance

In a normal load condition and a heavy load condition.

- Performance at bay level:
  - Time to change status at LCD since change in position of a switchgear (in normal load condition):

Switchgear	*Test case*	*Test result (time in seconds)*
For example: H2145	1	
	2	
	3	

  - Time to change status at LCD since change in position of a switchgear (in heavy load condition):

Switchgear	*Test case*	*Test result (time in seconds)*
For example: H2145	1	
	2	
	3	

- Performance at station level:
  - Time to open screens (in normal load condition):

Screen	*Test case*	*Test result (time in seconds)*
For example: Single-line diagram of 230 kV	1	
	2	
	3	

- Time to open screens (in heavy load condition):

Screen	*Test case*	*Test result (time in seconds)*
For example: Single-line diagram of 230 kV	1	
	2	
	3	

- Time to change status at HMI after change in position of a switchgear (in normal load condition):

Switchgear	*Test case*	*Test result (time in seconds)*
For example: H2145	1	
	2	
	3	

- Time to change status at HMI after change in position of a switchgear (in heavy load condition):

Switchgear	*Test case*	*Test result (time in seconds)*
For example: H2145	1	
	2	
	3	

- Time to execute a control order from the HMI (in normal load condition):

Switchgear	*Test case*	*Test result (time in seconds)*
For example: H2145	1	
	2	
	3	

- Time to execute a control order from the HMI (in heavy load condition):

Switchgear	*Test case*	*Test result (time in seconds)*
For example: H2145	1	
	2	
	3	

- Performance at remote control level (using a master unit emulator):
  - Time to change status at the master unit emulator screen after a change in position of a switchgear (in normal load condition):

Switchgear	*Test case*	*Test result (time in seconds)*
For example: H2145	1	
	2	
	3	

  - Time to change status at the master unit emulator screen after change in position of a switchgear (in heavy load condition):

Switchgear	*Test case*	*Test result (time in seconds)*
For example: H2145	1	
	2	
	3	

13.8.3 Test of the Time Synchronization Mechanism

- Date and time actualize after a disconnection: OK ___ Not OK ___
- Confirmation of time accuracy: OK ___ Not OK ___

13.8.4 Test of Event Buffer Capability

- Storage events are transmitted: OK ___ Not OK ___.

13.8.5 Interlocking Logics

- Predefined interlocking is fulfilled: OK ___ Not OK ___.

13.8.6 Synchronization Features

Predefined scenarios are fulfilled: OK ___ Not OK ___.

13.8.7 Operational Logic of Transfer Switch

- Different conditions are fulfilled: OK ___ Not OK ___.

13.8.8 Tests on the Communication Link for Technical Service

- A predefined service is carry out successfully: OK ___ Not OK ___.

Table 13.1 Examples of nonstructured FATs

Action	Checking
Turn off one of the redundant power sources of any IED.	Check if an alarm appears, review the alarm text and look at any other system response.
Switch off any MCB of power distribution circuit.	Look at the system response.
Check whether any device/arrangement may require a unforeseen manual/automatic selector.	
Unplug any optic connection belonging to the station bus.	Check if an alarm appears.
Disconnect the GPS antenna.	Look at the system response.
Unplug the connection with remote control level.	Look at the system response.
Unplug a connection at the station LAN	Check if an alarm appears.
Simultaneously push several keys on the HMI keyboard.	Look at the system response.
Force a switchover between redundant units.	Look at the system response.
Try to make simultaneous switching operations from different control levels.	Look at the system response.

13.9 Nonstructured FATs

Beyond exploring device behavior, nonstructured FATs are addressed at simulating situations that may happen in real life during SAS operation stages. This part of the testing program has an open scope to give the SAS buyer the possibility to detect system weaknesses ahead of unexpected events occurring at the site. Although it is recommended that a test list is completed once the test arrangement is inspected and is ready to start the complete FAT program, a list of possible examples is indicated in Table 13.1.

13.10 After FATs

Although SAS vendors are diligent offering a well finished system sample when they call the client to carry out the FATs, often an After Facts test is needed to ensure that unsatisfactory items detected during the FATs have been solved. The scope of this extra test program depends on specific FAT results obtained in each particular SAS project.

Further Reading

CIGRE (2002) Acceptance testing of digital control systems for HV substations, *Electra Review* 201, 21–31, April 2002.

Obrist M., Gerspach, S. and Brand, K-P. (August 25–31, 2013) Acceptance, commissioning and field testing for protection and automation systems, *CIGRE Colloquium*, Paper 116, Belo Horizonte.

Parapurath, A.A., Gopalakrishnan, A., Krishhnamurthy, P. and Rajagopal, N. (2012) *Method and a System for Simulation in a Substation*, Patent Application Publication US 2012/0239373 A1, Sep. 20, 2012.

14

Commissioning Process

The commissioning process on new Substation Automation Systems is a critical stage for ascertaining that a system is installed correctly and that it will remain satisfactorily in service for its expected life-cycle. Such a process is addressed at confirming the following:

- The system was configured/integrated correctly.
- The components did not suffer damage during transportation.
- The components are assembled and connected correctly.
- The overall system correctly fulfills the functionality agreed in the contract between user and supplier.

This process should be identified as a relevant stage requiring its own planning, scheduling, management and monitoring during the SAS construction phase. A commissioning team should be assembled including representatives of all involved parties including the following as a minimum:

- Utility/substation owner/project manager
- Utility operations personnel
- System designer
- Main contractor
- Electrical subcontractor
- System integrator
- Erection subcontractor
- Vendor of main devices.

Roles and responsibilities of the various parties within the commissioning process may vary with the type of substation owner and the size of the system. In some cases, an independent

Substation Automation Systems: Design and Implementation, First Edition. Evelio Padilla.

engineer may be brought in to coordinate and oversee the commissioning process. If an independent engineer is not in charge, the design engineer should be retained to oversee the commissioning process. In either case, certain key components of the program that must be included for a successful outcome are the following:

- A commissioning plan containing a clear definition of the complete process and the parties, as well as roles and responsibilities of involved representatives.
- Integration of commissioning activities into the overall project schedule.
- An organized system of commissioning documentation.
- Development of written testing and verification procedures for every critical aspect of system performance.
- Review of these procedures by all affected parties prior to testing.
- Clear definition prior to testing of the criteria for test result acceptance.
- Procedures for correction and retesting in the case of failure.

As a previous step, once the installation activities are completed, it is recommended carry out an inspection on individual components and the overall system to verify the following:

- Nameplate details according to approved drawings.
- Any physical damage or defects and cleanliness.
- Clamps and connections.
- Condition of accessories and their completeness.
- Electrical clearance.
- Earthing connections.
- Correctness of installation with respect to manufacturer manuals or approved drawings.
- Correctness and condition of connections.

Following that inspection, a loop check is also recommended, in such a way that the wiring of each control loop is physically verified from the terminal box of the primary equipment to the bay controller or other IED. Cable, conductor, terminal points and terminal designations should be verified and marked off as such on a copy of the loop diagram or equivalent schematic or wiring diagram. Verification should be made by signal tracing and continuity verification. Tags and labels placed during installation should not be considered adequate verification.

An outline of different items to be considered in preparation of commissioning tests (also called site tests) and also an approach of test scope include the elements in the following sections.

14.1 Hardware Description

This refers to an available table containing the following information on all system components:

- Component name
- Model
- Serial number

- Manufacturer
- Function/position in the system.

14.2 Software Identification

This is a list of all software installed in the system components:

- Software name
- Software version
- Device in which the software is installed.

14.3 Test Instruments

A list of instruments and accessories to be used in the tests shall include:

- Instrument name
- Main functions
- Model
- Manufacturer
- Serial number.

14.4 Required Documentation

A series of documents shall be available at the test site. These documents include:

- Test program showing the test scope in chronological order.
- Test procedures giving a clear understanding of the tests to be done.
- Test protocol (blank format).
- Technical manuals of all system components.
- Software documents.
- Manuals of testing and engineering tools.
- Instruction manual of test instruments.
- A set of relevant drawings including those related to system overview, schematic diagram and single-line diagram of the substation.

14.5 Engineering Tools

All engineering tools needed for system engineering and maintenance must be available and identified before starting the tests.

14.6 Spare Parts

Available spare parts must be duly identified.

14.7 Planned Commissioning Tests

14.7.1 System Start-Up

- Operative system start-up: OK ___
- User administrator feature works right: OK ___
- SAS main application start-up: OK ___

14.7.2 Displaying and Exploring the Main Menu Screen

- Single-line diagram of first primary voltage appears: OK ___
- Single-line diagram of second primary voltage appears: OK ___
- Single-line diagram of auxiliary power system appears: OK ___
- SAS scheme screen appears: OK ___
- Report screens appear: OK ___
- Alarm list appears: OK ___
- Event list appears: OK ___
- Contents from clicking optional keys are acceptable: OK ___

14.7.3 Displaying and Dealing with Single-Line Diagrams

- First primary voltage (e.g., 230 kV):
  - General appearance: OK ___
  - Bay/feeder arrangement and designations: OK ___
  - Busbar colors in different conditions: OK ___
  - Symbols of HV apparatus in different conditions: OK ___
  - Location of measured values: OK ___
  - Resolution of measured values: OK ___
  - Scales and units of measured values: OK ___
  - Indication of hierarchical control levels (L/R): OK ___
  - Control dialog boxes appear on request: OK ___
  - Opening commands progress if conditions are meet: OK ___
  - Opening do not progress if conditions are not meet: OK ___
  - Closing commands progress if conditions are meet: OK ___
  - Closing do not progress if conditions are not meet: OK ___
  - Blocking signals are active when conditions are meet: OK ___
  - Select before execute principle is implemented: OK ___
  - Hierarchical control level works correctly: OK ___
  - Messages from selected switchgear are shown: OK ___
- Second primary voltage (e.g., 115 kV):
  - General appearance: OK ___
  - Bay/feeder arrangement and designations: OK ___
  - Busbar colors in different conditions: OK ___
  - Symbols of HV apparatus in different conditions: OK ___
  - Location of measured values: OK ___

  - Resolution of measured values: OK ___
  - Scales and units of measured values: OK ___
  - Indication of hierarchical control levels (L/R): OK ___
  - Control dialog boxes appear on request: OK ___
  - Opening commands progress if conditions are meet: OK ___
  - Opening do not progress if conditions are not meet: OK ___
  - Closing commands progress if conditions are meet: OK ___
  - Closing do not progress if conditions are not meet: OK ___
  - Blocking signals are active when conditions are meet: OK ___
  - Select before execute principle is implemented: OK ___
  - Hierarchical control level works correctly: OK ___
  - Messages from selected switchgear are shown: OK ___
- Auxiliary power system:
  - General appearance: OK ___
  - System arrangement and designations: OK ___
  - Busbar colors in different conditions: OK ___
  - Symbols of MV/LV apparatus in different conditions: OK ___
  - Location of measured values: OK ___
  - Resolution of measured values: OK ___
  - Scales and units of measured values: OK ___
  - Indication of hierarchical control levels (L/R): OK ___
  - Indication of manual/auto condition of transfer switch: OK ___
  - Control dialog boxes appear on request: OK ___
  - Opening commands progress if conditions are meet: OK ___
  - Opening do not progress if conditions are not meet: OK ___
  - Closing commands progress if conditions are meet: OK ___
  - Closing do not progress if conditions are not meet: OK ___
  - Blocking signals are active when conditions are meet: OK ___
  - Select before execute principle is implemented: OK ___
  - Hierarchical control level works correctly: OK ___
  - Interlocking logic on transfer switch works correctly: OK ___
  - Messages from selected switchgear are shown: OK ___

14.7.4 Displaying and Dealing with the SAS Scheme Screen

- General appearance: OK ___
- Devices arrangement and designations: OK ___
- Texts on manual/auto for redundant schemes: OK ___
- Texts readability: OK ___
- Alarm signals are active when condition are meet: OK ___
- Specific device screen appear on request: OK ___
- Redundant schemes work properly: OK ___
- Specific devices screens are acceptable: OK ___
- Device parameterization facility works: OK ___

14.7.5 Displaying and Dealing with Report Screens

- Active energy report:
  - General appearance: OK ___
  - Color presentations: OK ___
  - Shown values are correct: OK ___
  - Scales and units are appropriate: OK ___
  - Handling facilities are friendly: OK ___
- Reactive energy report:
  - General appearance: OK ___
  - Color presentations: OK ___
  - Shown values are correct: OK ___
  - Scales and units are appropriate: OK ___
  - Handling facilities are friendly: OK ___

14.7.6 Displaying and Dealing with Measurement Screens

- Measurement on first primary voltage side:
  - General appearance: OK ___
  - All pertinent equipment/feeders are included: OK ___
  - All needed parameters are indicated: OK ___
  - Measuring units are appropriate: OK ___
- Measurement on second primary voltage side:
  - General appearance: OK ___
  - All pertinent equipment/feeders are included: OK ___
  - All needed parameters are indicated: OK ___
  - Measuring units are appropriate: OK ___
- Measurement on auxiliary power system:
  - General appearance: OK ___
  - All pertinent equipment/feeders are included: OK ___
  - All needed parameters are indicated: OK ___
  - Measuring units are appropriate: OK ___

14.7.7 Displaying and Exploring the Alarm List Screen

- General appearance: OK ___
- Date and time appear: OK ___
- Switchgear/feeders are indicated: OK ___
- Alarm texts are comprehensible: OK ___
- Alarm states are shown: OK ___
- Color presentation is adequate: OK ___
- Acknowledge facility works: OK ___
- Scrolling facility works: OK ___
- Relationship with external alarm annunciator: OK ___

14.7.8 Displaying and Exploring the Event List Screen

- General appearance: OK ___
- Time stamping appear: OK ___
- Switchgear/feeder are indicated: OK ___
- Event texts are comprehensible: OK ___
- Events appear in chronological order: OK ___
- Scrolling facility works: OK ___

14.7.9 Checking Peripheral Components

- Event data logger:
    - Start-up process: OK ___
    - Presentation format is adequate: OK ___
    - Automatic copy mode works: OK ___
    - Automatic switchover to redundant unit works: OK ___
- Hardcopy printer:
    - Start-up process: OK ___
    - Presentation format is adequate: OK ___
- External alarm annunciator:
    - Start-up process: OK ___
    - The set of displayed alarms is correct: OK ___
    - The alarm acknowledgement facility works: OK ___
    - The relationship with sonorous alarm is correct: OK ___

14.7.10 Checking the Time Synchronization Mechanism

- Time setting of time server: OK ___
- Time synchronization on station controller: OK ___
- Time synchronization on bay controllers: OK ___

14.7.11 Testing Communication with the Remote Control Center

Sometimes this is done using a master unit emulator:

- Changes of binary inputs are shown on master emulator: OK ___
- Changes of measured values are shown on emulator: OK ___
- Control commands progress if conditions are meeting: OK ___
- Commands do not progress if conditions are not meet: OK ___
- All data from general interrogation is shown on emulator: OK ___

14.7.12 Checking System Performance

- In a normal load condition from the HMI:
    - Change in position less than a predefined figure: OK ___

    - Change of measured value less than a predefined figure: OK ___
    - Command execution faster than a predefined figure: OK ___
- In an avalanche condition from the HMI:
    - Change in position less than a predefined figure: OK ___
    - Change of measured value less than a predefined figure: OK ___
    - Command execution faster than a predefined figure: OK ___
- In a normal condition from the master unit emulator:
    - Change in position less than a predefined figure: OK ___
    - Change of measured value less than a predefined figure: OK ___
    - Command execution faster than a predefined figure: OK ___
- In an avalanche condition from the master unit emulator:
    - Change in position less than a predefined figure: OK ___
    - Change of measured value less than a predefined figure: OK ___
    - Command execution faster than a predefined figure: OK ___

14.7.13 Testing Functional Performance

Performance testing should be carried out to verify compliance with the specified sequences of operations and control diagrams. The test consists of executing written step-by-step procedures in which a condition is initiated or simulated and the response of the system is noted and compared to the expected response. Functional performance tests are addressed to verify the following:

- Manual and automatic control modes.
- Normal system condition and operation modes
- Effect of all control orders.
- Implementation of all interlocking and blocking conditions
- Confirmation of failure state of all system components.
- Physical and information security measures.

14.8 Nonstructured Commissioning Tests

Similar to the Factory Acceptance Tests mentioned in Chapter 13, nonstructured commissioning tests are addressed at simulating situations that may happen during the SAS operation stage. This part of the testing program shall have open scope to give the SAS user the possibility to detect system weaknesses ahead of unexpected events. A list of possible examples is indicated in Table 14.1.

14.9 List of Pending Points

Major unsatisfactory results detected in commissioning tests must be duly recorded in a pending point list to be solved later without affect testing program efficiency.

Table 14.1 Examples of nonstructured commissioning tests

Action	Checking
Turn off one of the redundant power sources belonging to main IEDs.	Check if an alarm appears, review the alarm text and look at any other system response.
Switch off MCBs of power distribution circuits.	Look at the system response.
Unplug optic connections belonging to the station bus.	Check if an alarm appears.
Disconnect the GPS antenna.	Look at the system response.
Unplug the connection with remote control center.	Look at the system response.
Unplug connections at the station LAN	Check if an alarm appears.
Push simultaneously several keys on the HMI keyboard.	Look at the system response.
Force switchover actions between redundant units in both HMI and bay controllers.	Look at the system response.
Tray to make simultaneous switching operations from different control levels.	Look at the system response.

14.10 Re-Commissioning

Whenever all parts of the system are finally modified, repaired or replaced, re-commissioning is required to verify that the portions of the system affected function correctly and that the work has not affected other portions of the system. The extent of re-commissioning required should be determined from the extent of modifications. Examples are indicated here:

- For work that affects only devices and wiring external to the bay controllers, the affected loop should be verified and functionally re-tested.
- For changes to bay controller program logic or setting, the entire subsystem supported by the controller should be functionally re-tested.
- More extensive modification may require re-commissioning of the complete system.

Further Reading

Dierks, A. (n.d.) *Challenges Facing Protective Relay Engineers in Modern Times*, Paper for Alectrix, Ltd.
Energon Energy, Test and Commissioning Manual, Report Reference MN000301R171.
Leoni, A.R. and Nelson, J.P. (2000) Some lessons learned from commissioning substation and medium voltage switchgear equipment, paper for NEL Electric Power Engineering, Inc.

15

Training Strategies for Power Utilities

At the beginning, when digital technology begun to be used in substation environments, utilities faced difficulties in the implementation phase due different factors such as the following:

- Reluctance of experienced technical staff to embrace the new ways of designing and implementing secondary systems.
- Lack of standardization between different systems.
- Conflicts between the tendency of function integration imposed by the new technologies and the existing corporative culture.
- Concerns on restrictions for future SAS extensions.
- The complexity of applicable international standards.

Nowadays, although some utilities are still accommodating their technical staff to deal efficiently with new situations, the majority of them have become progressively empowered with the new skills required to achieve success in SAS projects. However, there is still a long way for power utilities and other substation owners to go in order to reach a comfortable level of understanding and management of the current technology. Considering that training is the key task to reach such a goal, two training means are identified:

- Project-related training.
- Corporate training.

The scope outlining both training modalities is presented in the following sections.

Substation Automation Systems: Design and Implementation, First Edition. Evelio Padilla.

15.1 Project-Related Training

Most of the current utilities' capabilities in new technology was reached through training programs provided by SAS suppliers, essential to specific system comprehension, operation and maintenance. Due the importance of such training programs, a recommended scope is presented.

15.1.1 Station Level Module

This training module covers the overall composition and operative philosophy of the provided system including structure, functionalities and attributes. The goal is to learn how the system looks from the control desk located at the main control house and how it interacts with upstream and downstream control levels. The proposed scope includes at least the following topics:

- Concept and features of the system:
  - Basic definition and principles
  - System structure
  - Station LAN
  - Station bus
  - Process bus
  - Functionalities in control, monitoring, protection, measurements
- Control function from station level:
  - Control procedure
  - Process database
  - Application software
  - Programming language
  - HMI facilities
  - Interfaces to IEDs
- System engineering:
  - Introduction
  - Signal name conventions
  - Engineering tools introduction and applications
  - Picture functions
- Process data engineering:
  - Introduction
  - Data acquisition mechanisms
  - Engineering tools
- Communication with remote control center:
  - Physical media
  - Communication protocol
  - Protocol conversion
- Practical exercises:
  - Pictures exploration
  - Picture edition
  - Control command execution

  - Change of interlocking logic
  - Change of synchronization condition
  - Use of peripheral devices
- Handling and support facilities:
  - System alarms
  - Communication errors
  - Backup procedures.

15.1.2 Bay Level Module

Bay controllers and protective relays lodged in local control rooms are the executing arms of Substation Automation Systems (SASs). A good comprehension of their physical composition and functional capabilities is essential for operation and maintenance personnel. People belonging to other utility units may also gain benefit learning details useful for future SAS projects. The proposed training scope for this module includes the following items:

Bay controller

- Hardware/software concept:
  - Introduction
  - Physical composition
  - Device layout
  - Base software
  - Dataflow
  - Main features
  - Documentation
  - Technical data
  - Self-supervision
  - Interfaces
- Engineering proceeding:
  - Engineering tools
  - Function implementation
  - The menu structure
  - Libraries
- The user program theory and praxis:
  - Overview and concepts
  - Installation and start-up
  - Build-up of project structure
  - User interface
  - Event and alarm handling
- Practical exercises:
  - Configuration of BIO cards
  - Configuration of LCD
  - Interlocking configuration
  - Engineering of signal commands.

Protective relay (each main type)

- Hardware/software concept:
  - Introduction
  - Physical composition
  - Device layout
  - Dataflow
  - Main features
  - Documentation
  - Technical data
  - Self-supervision
  - Interfaces
- The user program theory and praxis:
  - Overview and concepts
  - Installation and start-up
  - Menu structure
  - User interface
  - Device configuration
  - System parameters
  - Event handling
  - Data display
- Practical exercises:
  - Selection of the device functionality
  - Connection, installation and start-up
  - Configuration and device parameterization
  - Data storage and shut-down
  - Testing and diagnosis.

Controller of the auxiliary power system

- Hardware/software concept:
  - Introduction
  - Physical composition
  - Device layout
  - Base software
  - Dataflow
  - Main features
  - Documentation
  - Technical data
  - Self-supervision
  - Interfaces

- Engineering proceeding:
  - Engineering tools
  - Function implementation
  - The menu structure
  - Libraries

- Practical exercises:
  - Configuration of LCD
  - Interlocking configuration
  - Engineering of signal commands
  - Change in operational logic of transfer-switch.

15.1.3 Process Level Module

This training module must be addressed to the segment of the SAS that constitutes the interface between the primary equipment installed at switchyard and control, and protection cubicles located in local control rooms. It consists of a set of merging units and switchgear drivers linked by a dedicated fiber-optic network. The recommended training scope includes:

- Hardware/software concepts:
  - Introduction
  - Physical compositions
  - Device layouts
  - Functionalities
  - Base software
  - Dataflow
  - Concept of process bus
  - Main features
  - Documentation
  - Technical data
  - Self-supervision
  - Interfaces
- Practical exercises:
  - Device configurations
  - Testing and diagnosis.

15.2 Corporate Training

Further to specific knowledge on already installed SASs, utilities and other substation owner needs to conform and retain a strong background related to the concepts and philosophies applied nowadays in design and construction of secondary systems. Due the importance of such requirements, a proposed training scope is presented in the following subsections.

15.2.1 General Purpose Knowledge

Although by nature, power engineering expertise remains the core matter in which power utilities need to have major strength, it is now also mandatory that utilities staff refresh and keep updated on a minimum knowledge about latest technical developments and trends in the field of modern SASs. The best approach to achieving such a goal is by attending regular training activities offered by vendor-independent training providers in which at least the following topics are covered:

- Distributed control theory:
  - Features
  - Comparison with centralized solutions
  - Advantages/benefits
- Substation concepts:
  - Primary and secondary equipment
  - Control principles
  - Monitoring principles
  - Signalizing principles
  - Synchrocheck principles
  - Interlocking principles
  - Time synchronization principles
  - Importance of alarms and events display
  - Time stamping
  - Measurement principles
  - Metering principle
- Control and monitoring needs into substation environment:
  - Actions required for power system operation
  - Visualization of operative conditions of switchgear
  - Display and recording of power transformer parameters.
- Current involving of digital technology:
  - Types of signals
  - Data communication principles
  - Software applications
- Communication networks:
  - Configuration options
  - Quality attributes
- Ethernet networks:
  - Features
  - Advantages/benefits
- Intelligent Electronic Devices:
  - Basic principles
  - Typical composition
  - Functionalities
- The standard IEC 61850:
  - Overview on different parts
  - Main goals and referred issues
  - Communication services
  - Abstract models
  - Engineering process
- Scada systems:
  - Principles and purpose
  - Communication philosophies
  - Communication infrastructure

- Communication protocols:
  - Types
  - Functionalities
  - Control and data flow
- DNP3 protocol:
  - Main features
  - Advantages/benefits
  - Implementation
- Overview of typical systems:
  - Network topology
  - Devices identification
  - Redundancies
  - Application software
  - Expected performance.

15.2.2 Learning from the Standard IEC 61850

In comparison with technical standards dedicated to other substation components, Standard IEC 61850 looks hard to understand by power utility staff. This was the reason for finding the insufficient statement of: "Communication shall be implemented according to the standard IEC 61850" in many SAS specifications. Over time, more and more utilities have understood that they can not just close their eyes and expect that multiple IED-vendors involved in one SAS project should install interoperable systems. This positive view on the subject converts the learning on Standard IEC 61850 to a real challenge that needs to be seriously faced up to. In support of such a goal, it is recommended firstly that utility personnel get access to all relevant parts of the standard and go into them closely. Then, having gained knowledge on basic principles, get to a good vendor-independent training provider to discuss, at the very least, the following items:

- Parts of the standard:
  - Identification
  - Purpose
  - Content
  - Applicability
- General requirements established for SAS:
  - Reliability
  - Availability
  - Security
  - Environmental conditions
  - Concept of interoperability
- Standardized SAS structure:
  - Control levels
  - Logical interfaces
  - Concept of IEDs
  - Concept of the station bus
  - Concept of the process bus

  - Function allocation principles
  - Concept and application of the logical node
  - Function description
- Communication services defined by the Standard:
  - MMS service
  - GOOSE messaging service
  - Sampled values service
  - Application of different services
- Function implementation:
  - Control function
  - Monitoring function
  - Control constrains
- Engineering process:
  - Engineering concept
  - Configuration description language
  - SCL files
  - Engineering tools
- Conformance testing:
  - Tests scope, purpose
  - Test procedures
  - Quality assurance in testing execution
  - Test tools.

15.2.3 Dealing with Engineering Tools

Power utilities have a key role in the cooperative goal of getting interoperable systems. It consists of defining their substation specification using specific SCL files in line with IEC 61850 rules. These files are generated and managed through dedicated engineering tools. Further, power utilities staff also need to interact with various IEDs supplied under different SAS projects for parameterization and maintenance purposes. Because of this, the following training scope on that respect is recommended:

- Definition of substation specifications:
  - Evaluation of vendor-independent engineering tools
  - Managing of device libraries
  - Concept and application of logical nodes
  - Implementation of logic connections
  - Fixing of functional relations
  - Mapping of logical functions
  - Generation of SSD file.
- Managing of proprietary engineering tools:
  - Checking and fixing system requirements
  - Installation procedures
  - License set-up
  - Use of the tool.

Further Reading

Atlan Engineering 61850 (February 2014) Atlan Spec Users Manual Version V1.0.

CIGRE (December 1999) Power system operator training program design, development and utilization *Electra Review* 187, 117–131.

PullNet (May 2010) *Advanced Tools for Automated Substation Design*, white paper.

SENA technologies (2002) Introduction to Serial Communication, Technical Tutorial.

Wester, C. and Adamiak, M. Practical Applications of Ethernet in Substations and Industrial Facilities, Paper for GE Digital Energy Multilin.

16

Planning and Development of SAS Projects

In a power systems context, control consists of monitoring the states of voltages, currents and frequencies as critical parameters to keep the system stable and safe. This brings about the need for detecting when those parameters vary from the desired states for taking actions to restore them. Such control is made in a discrete way through both manual and automatic modes on a continuous basis. The variables are sensed by using instrument transformers that form part of the substation primary equipment. The control actions, which open or close primary switchgear, are delivered by control and protective relays belonging to the Substation Automation System (SAS). This system as a whole constitutes a combination of resources in the form of devices, software, materials, communication infrastructures, data and facilities, integrated in such a manner as to accomplish a set of functionalities addressed to preserve integrity of the power system and keep the electrical power flow safe and reliable. Similar to other electrical systems, SAS projects include the following stages:

- System specification
- Contracting process
- Definition of the definitive solution
- System design and engineering
- System integration
- Factory acceptance tests
- Site installation
- Commissioning process.

Each of these project stages contains key milestones that must be met and issues that must be addressed and resolved before moving on to the next stage.

A summary of the scope and detail of each stage is presented here in a way that can serve as a checklist and procedure guide for the SAS project lead.

Substation Automation Systems: Design and Implementation, First Edition. Evelio Padilla.

16.1 System Specification

The initial SAS requirements are outlined by the substation owner through a combination of formal specification and planning information. Specification covers the technical aspects to be considered for system design while planning information refers to all management premises to be followed to fulfill the project objectives.

Commonly, the scope and depth of SAS technical specification varies significantly from a substation owner to another. In extreme cases, for example when the substation owner provides just a single-line diagram and few sentences to describe the expected SAS, there is a high risk of affecting project development due the possibility of discovering later what is really required. Other extreme cases exhibit an exaggerated level of detail, sometimes including the desired SAS topology. These cases may force SAS suppliers to offer unproven prototype solutions.

A reasonable intermediate detail level of the specification may be achieved establishing in the specification using the following items:

- General system description including distributed control levels and operative philosophy.
- Single-line diagram of primary equipment.
- Single-line diagram of auxiliary power system.
- Substation physical layout including main control house and local control rooms, location of primary and secondary equipment, as well as cable routing and distances.
- Control philosophy for different voltage levels.
- Protection philosophy for different protective schemes.
- Features and formats for reports.
- Alarm requirements.
- Event achieving and handling requirements.
- Outline of communication path to remote control center.
- Requirements for HMI.
- Requirements for interlocking logics.
- Criteria for supervision and monitoring of primary equipment.
- Quality parameters in terms of reliability and availability.
- Requirements for cyber-security
- Requirements for engineering tools.
- Applicable standards (IEC 61850 and others)
- Restrictions on the application degree of Standard IEC 61850 (if any will be imposed).
- Requirements for spare parts.
- Requirements for training programs.
- Requirements for personnel participation in the engineering process.
- Terms of required guarantee.
- Requested compromise for providing spare parts and technical support during the complete life-cycle of the system.

16.2 Contracting Process

It can be seen that a variety of contracting modes are applied around the world for SAS implementation. Some substation owners buy separate IEDs and materials to assemble their systems "in house" by using their own personnel. Other prefer apply a turn-key contracting mode basis, most of the time forming part of a complete substation project, in which a main contractor is in charge of all activities from design until commissioning. When this type of contracting process is applied, the next four steps are followed:

1. Distribution of call for bids, or Request For Proposal (RFP) to potential suppliers (by side of substation owner).
2. Bid preparation and submittal (from SAS vendors).
3. Bid evaluation and selection of supplier (from substation owner).
4. Award of contract (from substation owner).

The call for bids is a data package used as formal mechanism by which the substation owner specifies the expected SAS details. The activity of bid preparation requires significant effort from the vendor proposal team, often including some pre-design and development tasks, in order to fulfill instructions contained in the call for bids.

The work of bid evaluation is the core activity of contracting process, particularly when several bids meet the basic requirements established into the specification. Guidance from the IEC 61069 Standard series is recommended in this phase of the project.

To perform a well-supported evaluation, it is essential that all bids include at least the following items:

- Description of the offered system under the standard concept of the vendor including logical and physical topologies.
- Description of all changes or new features developed under the standard concept to fulfill the specification.
- Description of communication buses (station bus, station LAN and process bus).
- Communication protocols to be used for data management.
- Protocol converter, concept and detail.
- Methods of data handling and storage.
- Time synchronization system.
- Mechanism for time stamping.
- Measures to ensure data consistency.
- Means for integrating protective relays.
- Means for integrating the auxiliary power system.
- Time values that reflect system performance.
- Reserve capabilities.
- Possibilities for future extensions (related to hardware and software).
- Self-monitoring facilities.
- Availability calculation on the overall system.
- Demonstration of compliance with EMC regulations.
- Set of engineering tools, including manuals and application purpose on the specific offered system.
- Data sheet of all system components.
- Software programs information including manuals, functions and devices in which they will be loaded.
- Peripheral components including data logger and external alarm annunciator, description and main features.

- HMI, hardware and software composition and details.
- Requirements of power supply for different sets of components including HMI, bay controllers and auxiliary elements.
- A proposed training program for substation owner personnel.
- A preliminary scope of Factory Acceptance Testing.
- A preliminary scope for Site Acceptance Testing.
- List of recommended spare parts.
- List of past projects in which the offered system had been installed.

Beside technical aspects, in the bid evaluation phase the substation owner is often concerned about buying equipment that will be affected by early obsolescence, as well as relating to the real possibility that the system will have to be expanded in future by using components coming from other SAS suppliers.

16.3 Definition of the Definitive Solution

Once the contract has been awarded, the opportunity to improve the selected system and clarify all related details appears before a formal contractual arrangement is signed between substation owner and chosen supplier. Typical modifications realized in this stage include the implementation of redundant schemes, confirmation of required hardware considering real amounts of signals and the number of bays/feeders to be assigned per bay controller, as well as replacement of certain components. Clarifications are addressed mainly to ensure that the supplier has a high degree of comprehension of the system specification.

16.4 Design and Engineering

The goal of the design and engineering stage is to translate the contractual design basis into a system design and documentation that is clear and complete so it can be assembled properly, tested and commissioned, as well as operated and maintained reliably and efficiently. Essential criteria for design documents are completeness in detail and clarity of the intended design. The system must be fully defined during this stage, where careful review can take place, rather than leaving details to be solved in the field during the installation or commissioning stage.

Detailed design documents must be formally reviewed by the substation owner representatives during the design stage through successive review submittals depending of the complexity of the document subject. Further to the specific technical aspects, the reviewers must verify opportunely the following issues:

- A complete list of design documents has been agreed.
- All symbols, abbreviations and line types are clearly defined.
- A complete and consistent cabinet identification scheme is used.
- All support documentation coming from third parties is available.
- The system adopted as the definitive solution has been analyzed using reliability, availability and maintainability criteria.
- Power supplies match with redundancy criteria.
- The auxiliary power system is able to provide the required power.

- Physical segregation exists between redundant components and paths.
- Separate power circuits source redundant components.
- Layout of operator controls and indicators and HMI screens has been clearly based on ergonomic considerations.
- Environmental protection level of devices and enclosures is clearly considered for their location.
- The designs of earthing and surge protection systems are appropriate.
- External communication links and interfaces are clearly defined.
- The design includes all testing and measuring provisions needed for commissioning purpose.
- All components can be tested and maintained without interruption to system functionalities.
- Physical security provisions are clearly defined.

16.5 System Integration

System integration is performed by either the SAS supplier or an external system integrator. This stage comprises the following activities:

- Overall configuration of databases.
- Component assembly.
- Device and system configuration: Based on user specification, the desired functions are allocated into IEDs by using specific configuration tools. Then communication infrastructure is fixed with data exchange criteria.
- Device parameterization: Refers to deployment of setting parameters and operating codes to the IEDs.
- Testing and diagnosis: A set of tests are realized with the support of testing and diagnosis tools for confirm proper operation of the system.
- Identification of failure modes as a baseline for trouble-shooting purposes.

16.6 Factory Acceptance Tests

This stage of the project is the ideal time to appreciate strengths and weaknesses of designed system. Some experimented non-complaint test results include:

- Loss of events when a switchover takes place in redundant schemes.
- Events do not appear in chronological order on the respective list.
- Difficulties in transmitting time stamping to the remote control center.

FAT preparation merits careful attention to ensure efficient development in execution. Before starting the tests, the following items must be ensured:

- A well understood document containing test procedures and acceptance criteria is available.
- The method to simulate the non-existent part of the system has been agreed.
- All alarm and event texts are written in the autochthonous user's language.
- The means to create a heavy load condition in network traffic is clearly defined.

- The way to determine the load condition in the CPU is clearly defined.
- Facilities to simulate a master control unit are available.

16.7 Site Installation

In this stage all system components are placed in compliance with the design documents. This is realized by the application of standard construction and management practices and techniques to handle schedule, cost and contract changes. However, in line with the nature of the specific project, the following additional previsions are recommended for a successful outcome:

- A change management process should be established assuring that field issues receive an appropriate technical review.
- All changes should be documented on an ongoing basis, and design documents kept current, rather than leaving this to an end-of-project "as-built drawings preparation" task.
- Operation and maintenance manuals should be assembled and reviewed during the installation process and must be completed prior to commissioning activities.

16.8 Commissioning Process

The objective of the commissioning stage is to obtain formal verification that the installed system complies with and performs in accordance with the design intent as defined in the contractual documents. Key aspects of the commissioning process include:

- Preparation of a commissioning plan.
- Establishing the commissioning team.
- Defining commissioning resources.
- Preparation and approval of test procedures.
- Preparation and review of test protocols.
- Individual tests on devices and accessories.
- Integrity testing of installed devices and interfaces.
- Circuit verification.
- Functional performance testing.
- Site acceptance tests on the system as a whole.
- Preparation of a pending point list.
- Verification of complete system documentation.
- Preparation of a program to evaluate and solve pending points.

The commissioning team should include representatives from all involved parties, for example:

- Operative personnel of the substation owner.
- Design engineers.
- Main contractors.
- Electrical sub-contractors.
- System integrators.
- Installation sub-contractors.
- Device vendors.

The following essential items of the commissioning program must be considered:

- A clear definition of the process and the parties, roles and responsibilities (commissioning plan).
- Integration of commissioning activities into the overall project schedule.
- An organized system of commissioning documentation.
- Developing of written testing and verification procedures for every critical aspect of the system performance.
- Review of these procedures by all involved parts prior to testing.
- Clear definitions prior to testing of the criteria for test result acceptance.
- Procedures for corrections and retesting in the case of failures.

The integrity testing scope should include checking the performance of fiber-optical links. Also, electrical conductors should be tested for continuity and insulation resistance according to the industry standard for their voltage rating. The wiring of each control circuit should be physically verified from the terminals of primary HV apparatus to the bay controllers and protective relays. Cable, conductor, terminal cubicle and terminal designations should be verified and marked off as such on a copy of the wiring diagram delivered at the design stage.

Functional performance tests are addressed to verify compliance with the specified sequences of operation and control diagrams. The tests consist of executing the written step-by-step procedures in which a condition is initiated or simulated and the response of the system is noted and compared to the specified response. The scope of the tests must cover:

- Performance in local and remote control hierarchy.
- Manual and automatic control mode.
- Normal system conditions and modes of operation.
- Device performance under contingency conditions.
- Effect of all control facilities and accessories.
- Right implementation of interlocking and blocking logics.
- Confirmation of a failure state of all outputs.

It is important to note that during the commissioning stage, the person in charge of commissioning process is responsible for the verification of the entire system in accordance with the approved design. When the commissioning engineer detects issues far removed from the approved design, he must raise these issues with the appropriate subject matter expert for design validation to ensure that all design principles are respected and all components are adequate for their intended purpose.

16.9 Project Management

The complexity of SAS projects is due mainly to their multidisciplinary nature and this calls for a well structured project team conformed by the following participants:

- Project manager (a challenging position, recognized champions are suitable candidates)
- Executive of the substation owner
- Main contractor representative
- Operation and maintenance representative

- Control and protection specialists
- System designer
- SCADA system specialist
- Information security specialist
- Reliability analyst
- Commissioning expert.

Once the project is activated, an initiation meeting must happen to assemble the team, focus them on the objectives of the project as a common mission and put in place the resources necessary for them to accomplish that mission. Specific activities that must be accomplished for the project to move forward effectively include the following:

- Define the project scope and schedule.
- Establish the communication process to be followed between different team members.
- Establish a document management system.
- Establish a process for tracking pending issues and documenting their resolutions.

It is recommended that all project team members stay in the project until the contracted scope of the work has been completed.

16.10 Security Issues

Beyond good designed system and components, there are a lot of site-related aspects that merit attention in order to reach the goal of maintaining a safe system throughout its entire life-cycle. These aspects include those in the following subsections.

16.10.1 Environmental Security

Environmental considerations covers:

- *Protection against earthquake effects*: This refers to installation measures to ensure that devices and associated communications networks will maintain appropriate performance even during and after earth motion. Special care must be taken regarding base fixing of control cubicles and loose fitting terminations of long fiber-optic links.
- *Avoid contaminating agents*: SAS components must be protected from the effects of dust, dirt, water, corrosive agents or other fluid and contamination by appropriate location within the main control house and local control rooms. For outdoor cubicles, an early factory inspection will ensure that enclosures fulfill the specified degree of environmental protection. Care should be also taken to ensure installation methods and conduit and tubing penetrations do not compromise enclosure integrity.
- *Fire protection*: Main control house and local control rooms should be provided with a dry agent fire protection system.
- *Complementary protections*: Even if cubicle enclosure satisfies required environmental protection protocols, where exceptional extreme conditions could represent a significant threat, alternate means must be provided to protect the component from environmental contaminants.

16.10.2 Electromagnetic Security

Electromagnetic threats to SAS components include transient voltages, radio-frequency interferences, earth potential rises and magnetic impulses. Although IEDs are usually well designed against such threats, good practice in installation is also recommended to minimize the possibility of undesirable effects. In that sense, the following considerations are presented:

- *Transient voltages*: Electronic devices are vulnerable to transient voltages coupled through the control room from atmospheric effects (thunderstorms, lightning), incoming switching surges and switching of capacitive or inductive loads into the substation. This calls for evaluation of the use of transient voltage suppressors on the power supply circuits.
- *Radio-frequency signals*: In some cases, any device lodged into control room may act as a radio-frequency source that may interfere with other IED operation due to coupling through control circuits. Further, other ambient sources of radio-frequency signals may also exist including commercial signals and radiated signals coming from the switchyard. Measures addressed to prevent this type of interference include the use of a shielded twisted pair or twisted triple conductor for conducting low level signals, as well as wiring installation in continuous metallic conduit systems.
- *Earth potential rises*: Effects of this event are mitigated by proper connection of the device to the unified earth subsystem installed within the control room.
- *Magnetic impulse protection*: Requires magnetically continuous ferrous shielding that is often not provided by the enclosures of typical control devices. For this reason, most devices are vulnerable to magnetic impulses and must be protected by location and/or external shielding.

16.10.3 Physical Security

In general, SAS components are usually installed inside secure areas with the same or higher degree of security that is deemed appropriate for the primary equipment. However, the electronic nature of these systems provides opportunities that could be compromised from both inside and outside the secured area, which must be addressed, for example taking care of the following:

- Implementing a short time setting on password-protected screen access.
- Keeping different enclosures, like switchgear control boxes, and junction boxes installed in the switchyard, locked or secured with tamper-resistant hardware, as well as providing the doors in the building with tamperproof switches or other means for detecting attempted intrusion, all connected to a site security system.

16.10.4 Information Security

Connections from the SAS to other networks extending beyond the substation perimeter introduce the threat of attacks of various types, for example the following:

- Unauthorized network access (hacking)
- Capture and recording of transmitted data

- Data interception, alteration/re-transmission
- Replay of intercepted and recorded data
- Denial of system service.

The best defense against these threats is to avoid network connections with other networks and keeping the strictly necessary connections physically disconnected for sporadic use, for example those dedicated to allowing technical service from a remote station belonging to the SAS provider.

16.10.5 Software Security

Considering the inherent complexity and exposure to intentional software damage that can occur in a modern digital SAS, it is good practice to implement a software management and documentation system (SMDS), especially in cases of important substations. The SMDS is software that resides on a dedicated computer monitoring all activities of the SAS. It allows the operator to do the following:

- Maintain a system-wide repository for historical storage of the application configuration files.
- Identify exactly who has modified a system configuration or application parameter, what they changed, where they changed it from and when the change was made.
- Support application restoration following a catastrophic event.
- Generate views into the SMDS for more detailed analysis of configuration changes.

Additional benefits derived from having a SMDS include:

- Avoiding maintaining incorrect or incompatible software versions.
- Assuring there are not multiple versions of software on files.
- Preventing multiple users from causing a conflict somewhere on the system.
- Preventing legitimate changes from being reversed or overwritten.
- Supporting the availability of the system at its maximum.

16.11 Documentation and Change Control

Complete and accurate documentation is critical to the commissioning and ongoing maintenance and operation of SASs, and should be a high priority of the responsible design contract. The scope of drawing up a package to be provided by the system designer/integrator is indicated in Appendix B. Essential items include the following:

- One-line diagram
- Three-line diagram
- System topology
- Components data sheets
- Bay controller I/O wiring diagrams
- Signal lists

- Binary logic diagrams
- Control schematics
- HMI pictures and database
- Software configurations
- Means for physical and information security.

From this list, major complexity in preparation corresponds to control schematic drawings (one per bay). These drawings show the interface between the SAS and the internal wiring of related equipment such as circuit breakers, disconnectors, power transformers and other equipment and devices. Proper development of control schematics requires a significant effort on the part of the system integrator because the standard drawings provided by manufacturers of related equipment are typically inconsistent in format, symbols and contents, thus requiring that they be re-drafted by the system integrator to provide an integrated and comprehensible document.

Documents should be provided as both hard copy and appropriate electronic formats that are capable of having revisions made to them. All documents must include a creation date, issue date, revision date and revision history in a format that is consistent across all documents. A document database or spreadsheet must be maintained that provides a current listing of all project documents and their revision status. The database should include the following:

- Document type
- Document number
- Page number
- Title
- Current revision
- Revision date.

Each document submittal must include an updated submittal of the documents database. The database should be provided in an electronic format and maintained on an ongoing basis by the substation owner, even after system commissioning. All changes made to the system should be promptly reflected by revising the associated documentation, distributing copies of the revised documentation and updating the documents database.

Further Reading

CIGRE (April 2002) Acceptance testing of digital control systems for HV substations *Electra Review* 201, 21–31.

CIGRE (August 2002) Analysis and guidelines for testing numerical protection schemes *Electra Review* 191, 149–157.

Hirano, M., Nakagawa, N., Kimura, S. and Imamura, A. (1981) Project engineering and management for construction of substations on turn-key basis, *Hitachi Review*, 30, 3.

IEC 61069–1, Industrial-Process Measurement and Control – Evaluation of System Properties for the Purpose of System Assessment – Part 1: Evaluation of System Properties.

IEC 61069–2, Industrial-Process Measurement and Control – Evaluation of System Properties for the Purpose of System Assessment – Part 2: Assessment Methodology.

IEC 61069–3, Industrial-Process Measurement and Control – Evaluation of System Properties for the Purpose of System Assessment – Part 3: Assessment of System Functionality.

IEC 61069–4, Industrial-Process Measurement and Control – Evaluation of System Properties for the Purpose of System Assessment – Part 4: assessment of System Performance.
IEC 61069–5, Industrial-Process Measurement and Control – Evaluation of System Properties for the Purpose of System Assessment – Part 5: Assessment of System Dependability.
IEC 61069–6, Industrial-Process Measurement and Control – Evaluation of System Properties for the Purpose of System Assessment – Part 6: Assessment of System Operability.
IEC 61069–7, Industrial-Process Measurement and Control – Evaluation of System Properties for the Purpose of System Assessment – Part 7: Assessment of System Safety.
IEC 61069–8, Industrial-Process Measurement and Control – Evaluation of System Properties for the Purpose of System Assessment – Part 8: Assessment of Non-Task-Related System Properties.

17

Quality Management for SAS Projects

The meaning of *quality* is widely understood as "conformance to requirements". From such a sentence, it sounds easy achieve an acceptable level of quality if requirements are well identified and clearly stated. But it is not easy when so many actors are involved and the original system requirements are just lightly outlined, as often occurs in SAS projects. Because of that, a key activity in this type of project consists of building early, before starting the detailed engineering phase, a formal set of requirements by defining a basic system concept; for example, through the following four steps:

Step 1: Meeting for the basic concept definition. The aim of this meeting is to check thoroughly the range of delivery to detect possible missing or to eliminate double functions. The work-share must be discussed in detail and the work already done documented. It is recommended that the SAS supplier takes the initiative and leadership for this meeting, as well as the substation owner incorporating all the corresponding people at the first project meeting to ensure a dialog between involved engineers.

Step 2: Submission of examples and questionnaires. The SAS supplier will summit questionnaires to be filled in by personnel of the substation owner. Assistance may be required from the supplier in the filling out process to make sure the answers are as exact as possible.

Step 3: Elaboration of the basic concept. In this step, all parts of the desired system are worked out by the supplier according to buyer specification, complemented by inputs gained from the meeting and questionnaires. The last phase of this step should be an internal audit carried out by experts with wide experience in substation automation.

Step 4: Approval of the basic concept. In this step is very important to check carefully the features of the created basic concept and verify that all buyer ideas have been implemented correctly.

Substation Automation Systems: Design and Implementation, First Edition. Evelio Padilla.

Once buyer approval is obtained, the system concept must not be changed any more. The document containing the basic system concept must include, as a minimum, the following items:

- Introduction
- Project organization dates for work-share
- Hardware to conform station level
- Software to be loaded on different IEDs
- Sets of screens to be displayed
- Component codes and signal names
- Typical signal lists
- Protocol converter, concept and details
- Hardware to be placed on bay level, process connection, local displays
- Station bus, concept and details
- Tools, identification and applications
- Protection integration, concept and details
- Auxiliary power system, integration and details
- Hardware to be located on process level
- Process bus, concept and details
- Testing concept
- Commissioning concept.

17.1 Looking for Quality in Component Capabilities and Manufacturing

It is desirable that component manufacturing is based on the International Quality Assurance Standards ISO 9000 and ISO 9001, as the first step to achieving the overall goal of experimenting a maximum of two significant failures during the SAS's lifetime. In that respect, careful attention to issues of type and factory conformance tests is essential.

17.1.1 The Dilemma with Respect to Type Tests

What should be done with respect to type tests? This is a typical doubt from the side of the substation owner when an SAS project commences. A scenario in such a sense can be described by the following facts:

- SAS supplier provides type test certificates together with the bid.
- The reported tests were performed some years ago.
- There is the possibility of change and/or modifications between already tested devices and offered devices.
- Repeat type tests appear to be expensive.

It is common practice that the substation owner accepts certificates for type tests conducted in an accredited test laboratory according to international standards fulfilling the levels of

Table 17.1 Model of inspection and test plan ITP

Device subcomponent	Issue parameter	Type of action	Reference document

compliance specified. However, it is reasonable that the substation owner has the freedom to ask for repeats of any or all type tests at their expense based on a price list for the type tests included into the bid. Dielectric tests and protection against electromagnetic disturbance tests are recommended for repeating, particularly on the device that will be used as the bay controller.

17.1.2 The Importance of Factory Conformance Tests

These tests, performed on individual devices and sub-components, also called routine tests, are addressed at detecting manufacturing errors. The scope and procedure are indicated in the standard Inspection and Test Plan (ITP) of the manufacturer, which is usually structured as shown in Table 17.1. Early evaluation and discussion between SAS supplier and substation owner of the content of ITP is strongly recommended to ensure the test will accomplish affectively the assigned purpose.

17.2 Looking for Quality during the Engineering Stage

The SAS engineering process demands exact definitions of the functionality in an early project stage. This is the reason for creating the basic system concept in close cooperation between the substation owner and contractor before starting the detailed engineering phase. Having early a well-defined system is a significant advantage. However, there are also a number of issues that are equally important that can crop up at the beginning of the engineering process that could cause re-work and delays. Those issues include:

- *Project drawing list*: The SAS supplier shall submit a preliminary list for buyer comment and complementation. The list is often updated along the detailed engineering phase.
- *User requirements with respect to symbols and colors*: This applies for both screens and control cubicle sticks and buttons. In some cases, the user practice does not match with international standards.
- *Documents coming from third parties*: This refers to making all drawings available related to primary equipment; like power transformers, circuit breakers, disconnectors and other HV apparatus. These drawings are needed for the engineering team to assemble the system as a whole.

It is recommended to carry out a design review at the middle of the engineering stage with the participation of qualified personnel who can critically examine the work already done and plan for continuing work with pending engineering activities. Examples of items to check during such a design review cover the following.

General items:

- The number of bits to be used for position indication of disconnectors (ensure that the transit condition can be implemented).
- Details on interfaces for control and monitoring tasks, such as those associated with temperature in power transformers.
- Criteria for alarm grouping.
- Check that alarms generate their own event signals
- Measures in respect to cyber-security.
- Language used in text of alarms and events (ensure that it corresponds to the buyer's native language).
- Implementation of the select-before-operate principle in the control facility.
- Clear understanding of interlocking principles.
- Ensure that event messages to be transmitted to the remote control center will be duly time stamped.

Control cubicles (including those dedicated to the auxiliary power system):

- Installation of local/remote selectors.
- Check that local/remote condition is displayed on the LCD.
- Installation of locked/unlocked selector.
- Installation of cable entrance accessories.
- Robustness of door retainer.
- Identification of bay/feeder to which cubicles are dedicated.
- Identification of control buttons
- Identification of selectors for redundant control units.
- Identification of active control unit in cases of redundant arrangements.
- Identification of LEDs located on front plate of bay controllers.
- Cubicle colors
- Earthing facilities
- Mounting and fixing facilities
- Electrical connectivity between door and main body of the cubicle.

Control Desk:

- Dimensions
- Ergonomic structure
- Top protection.

HMI:

- Automatic start
- Pictures appearance and contents
- Alarm texts
- Event texts
- Control facility and procedure
- User administrator facility.

Factory Acceptance Tests (FATs):

- Discuss preliminary details in respect to test scope and procedures.

17.3 Looking for Quality in the Cubicle Assembly Stage

Often, the influence from the substation owner in this project stage is made by a factory inspection program carried out by either internal personnel or an independent inspection body. In line with applicable quality standards, the program covers the following three well defined quality phases:

- Incoming inspection: This inspection, made on components coming from sub-suppliers, usually includes visual control, dimensional control and checking of test certificates.
- Assembly line inspection: Refers to checking of mounting details like fixing, connections and soldering.
- Final inspection and tests: This inspection is made on the finished cubicle as shown in Figure 17.1, and must cover these aspects:
  - Cubicle identification.
  - Check that dimensions, presentation and device layout conform to the respective drawings approved by the substation owner.
  - Careful review of relay front plate to confirm that signals and colors fulfill buyer practice requirements.
  - Illumination devices, capacity and location.
  - Earthing terminal.
  - Identification of voltage outlets.

Figure 17.1 Control cubicle ready for inspection. Source: © Corpoelec. Reproduced with permission of Corpoelec

Table 17.2 Nonconformities detected in cubicle assembly stage

Nonconformity	Assessed significance		
	Low	Medium	High
A local/remote selector was missing in the control cubicle for the auxiliary power system		x	
Security keys for operating locked/unlocked selectors on control cubicles were missing			x
Some bay feeder names labeled on control cubicles were incorrect		x	
Control buttons were unidentified		x	
LEDs signalizing were unidentified		x	

- Cabling, fixing and arrangements.
- Packing provisions.
- Special measures for transportation.

Examples of nonconformities detected at this stage are listed in Table 17.2.

17.4 Looking for Quality during FAT

FAT should merit special attention from the substation owner. This is because the nature of the system means there are many details to check before the SAS components arrive on the site. A sample of nonconformities experienced in this stage is shown in Table 17.3.

17.5 Looking for Quality during Installation and Commissioning

As the cubicles and other components and accessories are installed, quality assurance procedures must be administered to verify components are installed in accordance with the minimum of manufacturer's recommendations, safety codes and acceptable installation practices. Quality assurance discrepancies must be identified and added to a "commissioning improving point list" that must be incorporated as part of the overall commissioning program. As a key issue, it is recommended throughout the commissioning process to keep a strict record of all maintenance relevant information, such as IED definitive firmware and configuration versions. Special care must be taken at this stage to avoid computer viruses spread via the use of portable drives/memory used by commissioning personnel.

Some nonconformities have been experienced at this stage and are indicated in Table 17.4.

17.6 Use of Appropriate Device Documentation

Unlike other equipment used in power systems, Intelligent Electronic Devices (IEDs) applied in SASs are documented in an increasingly sophisticated and segregated manner, as shown in Table 17.5. The use of the complete set of documents by the substation owner is very convenient.

Table 17.3 Non conformities detected in FAT

Nonconformity	Assessed significance		
	Low	Medium	High
Alarms were not generating their own events		x	
Blank lines were noted into event list	x		
Was lacking the event text related to the invalid disconnector position when fixed transit time is exceeded		x	
LEDs located in control units remain in red after the respective alarm has disappeared	x		
Signal of "selected equipment" remains after the timing for selection expires	x		
Selector invalid positions were not generating alarms		x	
The "disconnected" position of L/R/D selectors was generating alarms instead of just events	x		
The "invalid position" of switchgear was not generating alarms	x		
In the primary equipment screen, indications of local positions of respective selectors were missing	x		
Transit condition was shown for circuit breakers	x		
Diesel generator symbol was missing in the auxiliary power system picture		x	
A LV circuit breaker was labeled as a disconnector			x
Too much information shown in the screen related to a single-line diagram of the auxiliary power system	x		
A "local position" indication referred to a L/R selector that was incorrectly located in single-line diagram	x		
Incoherence in grouping of signals shown in HMI and local display of the controller for the auxiliary power system	x		
Position indications related to some LV circuit breakers were missing		x	
Identification of busbars belonging to primary circuits were missing in single-line diagrams		x	
Trip signal coming from MCB associated to line voltage transformer was blocking the control order on the primary switchgear		x	
A common alarm signal was implemented for failures of redundant LANs at station level		x	
Different timings were fixed in various control cubicles for selection times of switchgear	x		
It was possible to select another switchgear for command before selection time of the first selected switchgear expired		x	
It was detected that the unlocked/locked selector in the unlocked position and L/R selector in the local position made it impossible to take control from the bay control cubicles			x
It was possible to deliver a closing command on a circuit breaker while the associated disconnector was in the transit condition			x
Alarms related to the DC circuit dedicated to the bay controller faulty and bay controller faulty were missing		x	
Alarms related to failure in communication between redundant control units were missing		x	
Screen blinking was detected in certain conditions	x		

(*continued*)

Table 17.3 (*continued*)

Nonconformity	Assessed significance		
	Low	Medium	High
Commutation time too long between redundant control units	x		
Current switchgear statuses were not duly updated on the screen of the active control unit following a switchover		x	
Alarms referring to trip due high temperature in MV transformers do not appear on the screen		x	
Alarm texts differ from event texts		x	
Alarm referring to a lack of time synchronization signal was missing		x	
Alarms referring to the failure of the local display screen of bay controllers were missing		x	
Event signal referring to returning back to fiber-optic links to normal conditions were missing		x	
Alarm signals referring to failure of fiber-optic connections belonging to stand-by control units were missing		x	
Some erratic alarm messages appeared together with certain alarm signals	x		
Facility for fixing timing in automatic transfers of the auxiliary power system was not foreseen in the respective control unit		x	
Some messages displayed on the screens of control units were incorrect		x	
In some screens/diagrams keys/buttons to return to the main menu were missing	x		
Event message referring to the "out of synchronism" condition was missing		x	
It was possible make a control command from the control unit of auxiliary power system with the selector L/D/R of the distribution center in the "disconnected" position			x
It was necessary take some manual steps to start-up the HMI application		x	
Some event messages were incorrect	x		
It was possible to edit IP addresses in the "access to operator" control hierarchy		x	
Loss of information due HMI unit faulty was noted			x

Table 17.4 Nonconformities detected in the commissioning process

Nonconformity	Assessed significance		
	Low	Medium	High
Bay controllers show voltage and current values while no signal was applied in respective inputs		x	
Rated frequency of CT modules belonging to bay controllers was 50 Hz instead 60 Hz	x		
Internal logic of bay controllers was blocking control commands on switchgear (action allowed from the switchgear itself)			x
Internal logic of bay controllers includes trip order on circuit breakers when a pole discrepancy is detected (it should be made by the CB control circuit itself)		x	
In signaling the MCB condition opposite bits for control and protection relays (1 = normal condition for control and 0 = normal condition for protection) were used	x		

Table 17.4 (*continued*)

Nonconformity	Assessed significance		
	Low	Medium	High
Transit condition was implemented for circuit breakers	x		
Identification of which bay feeder is dedicated to which bay controller was not clear			x
Some event texts were incorrect or confusing	x		
Date format (YY/MM/DD) shown on HMI differs from native format	x		
Some texts on HMI pictures were in a foreign language	x		
Transit condition of disconnectors was implemented using a bit combination not corresponding with the standard of the substation owner		x	
Rated voltage of primary circuit was configured incorrectly	x		
Audible alarm was not implemented		x	
The switchgear selected condition remained in the main bay controller even after a switchover to the back-up controller had occurred	x		
Events associated to certain alarms were missing	x		
Some BIO cards belonging to bay controllers were not configured in the factory		x	
Some alarm and events texts differ from those indicated on project drawings	x		
Internal logic of bay controller blocks all switchgear belonging to the bay when any of the switchgear L/R selector is in the local position		x	
Fixed timeout for switchgear selection was too long	x		

Table 17.5 Set of documents referring to IEDs

Device document	Application ambit (stages of device life cycle)					
	Pur.	Eng.	Ins.	Com.	Ope.	Mai.
Buyer Guide	——					
Engineering Manual		——		——		——
Installation Manual			——	——		——
Commissioning Manual				——		——
Operation Manual				——	——	——
Service Manual						——
Application Manual	——	——				
Technical Manual	——	——	——	——	——	——
Communication Protocol Manual	——	——	——	——		——
Point List Manual	——	——	——	——		——

Notes:
Pur.: Purchase
Eng.: Engineering
Ins.: Installation
Com.: Commissioning
Ope.: Operation
Mai.: Maintenance

Further Reading

IEEE Std. 730, Standard for Software Quality Assurance Processes.
ISO 9001, Quality Management Systems.
Janssen, M.C. and Melenhorst, E. (Summer 2009) IEC 61850 and Quality Assurance, PAC.
Padilla, E. (1991) Performing factory inspections for special projects, *Power Technology International Review*, 117–119.
Pieters R., de Croon, J.A.W. and Sibbel, H.N.E. (n.d.) Quality Management for Engineering Projects for HV Installations, Paper for KEMA Transmission & Distribution.
Schneider Electric (2010) *Low Voltage Switchboards: Quality Inspection Guide*.

18

SAS Engineering Process According to Standard IEC 61850

The SAS engineering process, as defined by Standard IEC 61850, is the kingdom of computer files and programming tools. The IEC 61850–6 establishes a set of files under the rules of the Substation Configuration Language (SCL), which is based on eXtensible Markup Language (XML) that defines a set of principles for encoding documents in a format that is both human- and machine-readable. The Standard also establishes a group of engineering tools for creating/handling those SCL files.

18.1 SCL Files

An SCL file contains the following sections:

Header: This section is used to identify the version and other basic details of the SCL configuration file.

Substation: This is the section dealing with the different elements of the engineered substation, including various primary apparatus, HV connections and other facilities. The elements include voltage levels, bays, feeders, power transformers and switchgear like circuit breakers and disconnectors.

Communication: This section deals with different communication points (access points) for accessing the different IEDs of the complete system.

IED: This section describes the complete configuration of an Intelligent Electronic Device (IED). It contains different access points of the specific device, the logical format of the device, logical nodes and report control blocks coming under the specific IED.

Data-type templates: This section defines different IEDs in logical formats, logical nodes, data and other details separated into different instances. It includes also the complete data modeling according to IEC standards 61850–7-3 and 61850–7-4.

Substation Automation Systems: Design and Implementation, First Edition. Evelio Padilla.

Depending on the purpose of specific SCL file, it is classified into one of the following types:

1. IED Capability Description (*ICD*) file: Defines complete capability of an IED. This file needs to be supplied by each IED vendor to make the complete system configuration. The file contains a single IED section, an optional communication section and an optional substation section that denotes the physical entities corresponding to the IED.
2. System Specification Description (*SSD*) file: Contains the complete specification of a SAS including a single-line diagram for the substation and its functionalities (logical nodes). This will have a substation section, data type templates section and logical node type definitions, but need not have an IED section.
3. Substation Configuration Description (*SCD*) file: Describes the complete substation detail. It contains substation, communication, IED and data type template sections. An SSD file and different ICD files contribute to making an SCD file.
4. Configured IED Description (*CID*) file: The file used to make communication between an IED configuration tool and an IED. It can be considered an SCD file stripped down to what the concerned IED needs to know and contains a mandatory communication section of the addressed IED.
5. Instantiated IED Description (*IID*) file: Defines the configuration of one IED for a project and is used as data exchange format from the device configuration tool to the system configuration tool. This file contains only the data for the IED being configured: One IED section, the communication section with the IED's communication parameters, the IED's data type templates and, optionally, a substation section with the binding of functions (logical nodes) to the single-line diagram.
6. System Exchange Description (*SED*) file: This file is to be exchanged between the system configuration tools of different projects. It describes the interfaces of one project to be used by another project and, at re-import, the additionally engineered interface connections between the projects.

The last two file types were introduced with edition 2 of the Standard.

18.2 Engineering Tools

In the engineering process the following tools are used:

1. System Specification (*SST*) Tool, the purposes of which are the following:
   - Defining the primary circuit arrangement as indicated on single-line diagram provided by substation owner, including the process link for the secondary equipment.
   - Generating the System Specification Description file (SSD file).
2. System Configuration (*SCT*) Tool, the purposes of which are the following:
   - Importing the SSD file.
   - Importing IED Capability Description files (ICD files).
   - Configuring communication settings.
   - Configuring communication functions, including GOOSE messages.
   - Generating the Substation Configuration Description file (SCD file).
3. Device Configuration (*DCT*) Tool, the purposes of which are the following:
   - Importing the SCD file.
   - Configuring the device functions, such as control and monitoring functions.
   - Generating the Configured IED Description file (CID file).

- Generating the Device Parameterization file.
- Loading the Device Parameterization file into the target device.

18.3 Engineering Process[1]

Those resources, files and tools are applied in the engineering process following the flow diagram shown in Figure 18.1.

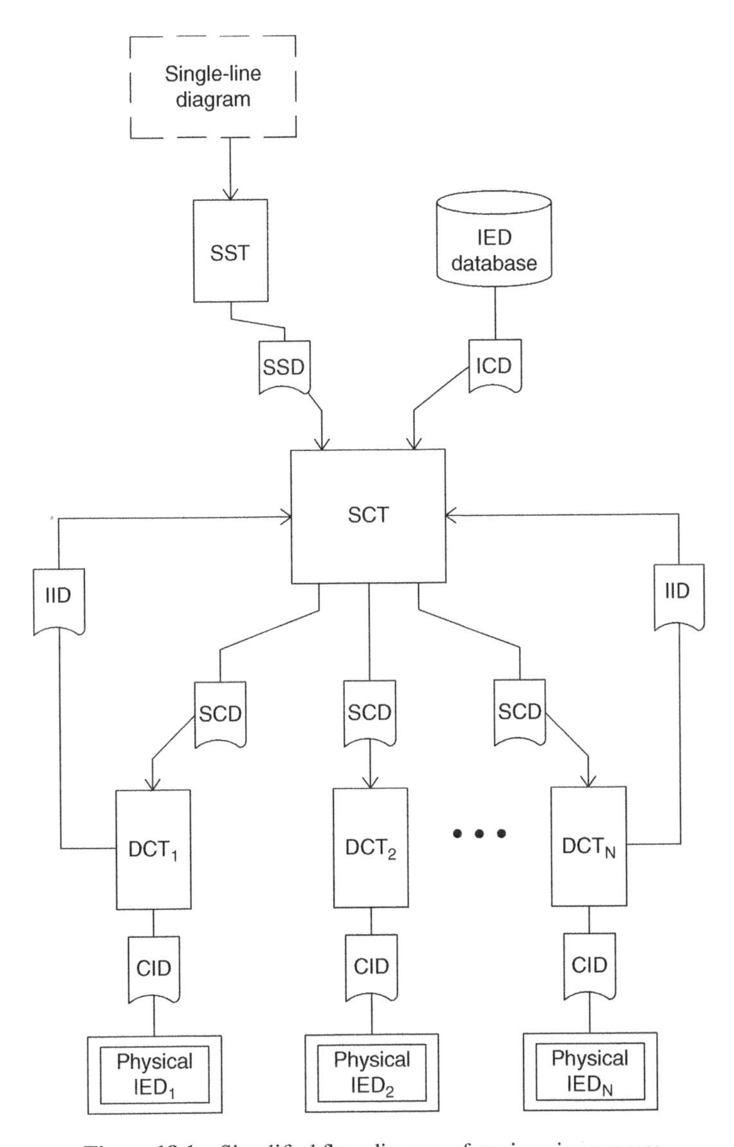

Figure 18.1 Simplified flow diagram of engineering process

[1] Adapted from VDE Association for Electrical, Electronic & Information Technologies (2009), DKE Working Group 952.0.1, http://www.vde.com/en/dke/DKEWork/NewsfromtheCommittees/Documents/e_AK95201_Engineering_Process_V10.pdf

The engineering process itself is developed through the following six steps:

Step 1: Definition of substation specification (by using the System Specification Tool: SST)

Task	*Input*	*Output*
Creating of function-related models	• Single-line diagram as provided by substation owner. • Set of control and protection functions needed. • Power system operative concepts.	• Set of requirements including; function tables of the SAS components, information model and logical diagrams for interlocking purpose.
Generation of the SSD file	Specification requirements	The SSD file

Step 2: Component selection

Task	*Input*	*Output*
Choosing of SAS components	• Component requirements including; function tables, details of connection with the remote control center, protection schemes, logical diagrams and functional criteria • SSD file.	• Prospective specification including; components number and layout and selected components. • Function allocation among different components. • ICD files.

Step 3: System definition

Task	*Input*	*Output*
Fixing of network topology	• Function tables of the SAS components • Information on selected components	• Network topology
Defining of the communications	• Requirements including; function tables of SAS components, logical diagrams, detail of connection with remote control center • Network topology.	• Communication parameters. • Detail for communication services, e.g., GOOSE messages.

Step 4: Components configuration (by using the Device Configuration Tool: DCT)

Task	*Input*	*Output*
Assignment of components functionalities	• Component capabilities. • Specification requirements.	• Functional • parameterization.
Generation of the CID file	• Specification requirements including; function tables, information model, details of connection with network control center and interlocking principles. • ICD file. • SSD file.	• CID file
Fixing of the communication features	• Network topology. • Parameters for communication services.	• Communication-related component configuration.

Step 5: System configuration (by using the System Configuration Tool: SCT)

Task	*Input*	*Output*
Interconnection of the substation section with the IED section into SCL file and parameterization the system configuration	• System communication parameters. • Network topology. • CID file • SSD file	• SCD file
Establishment of the information needed for tests and commissioning	• SCD file • Associated information such as addresses for connection with remote control center. • Wiring diagrams.	• Test specifications

Step 6: Definitive component parameterization (by using the Device Configuration Tool: DCT)

Task	*Input*	*Output*
Fixing the component parameterization	• Parameterization data referred to functionality • Parameterization data referred to communication. • SCD file.	• Particular component parameterization file.
Loading the parameterization data into the target component	• Particular component parameterization file	• Component parameterized

It has been recognized that by developing the SAS engineering process according to Standard IEC 61850, this is the basis for achieving significant time and cost reductions, changing the procedures for creating project documentation dramatically and enhancing the chance to obtain interoperability.

Further Reading

Apostolov, A. (2010) IEC 61850 Substation Configuration Language and Its Impact on the Engineering of Distribution Substation Systems, *CIDEL Conference*, Buenos Aires.

Atlan Engineering 61850 (February 2014) *Atlan Spec Users Manual* Version V1.0.

Brand, K.P., Brunner, C. and Wimmer, W. (2004) Design of IEC 61850 based substation automation systems according to customer requirements, paper B5–103, *CIGRE session*.

Bray, T., Paoli, J., Sperberg, C.M. et al. (November 2008) Extensive Markup Language (XML) 1.0, 5th edn, WC3 Recommendation.

Brunner, C., Reuter, J. and Hughes, R. (September 12–17, 2011) Choosing and using IEC 61850 SCL files, process and tools correctly throughout the complete SAS lifecycle, paper 104, *Study Committee B5 Colloquium*, Lausanne.

Brunner, C., Reuter, J., Muller, D., et al., Focus on the application – IEC 61850 experience with third party system configuration tool.

De Mesmaeker, I., Rietmann, P., Brand, K-P. and Reinhardt, P. (2005) Practical Considerations in Applying IEC 61850 for Protection and Substation Automation Systems, *GCC Power 2005 Conference and Exhibition*, Doha, Qatar.

Erik Herzog, E. (2004) *An Approach to Systems Engineering Tool Data Representation and Exchange, Linkoping Studies in Science and Technology*, Dissertation 867, Linkoping Universitet.

Gauci, A. (2013) *Effect on Substation Engineering Costs of IEC 61850 and System Configuration Tools*, Schneider Electric report.

Helinks STS (2015) Supplier of Vendor-Independent Engineering Tools, available at: www.helinks.com (accessed June 8, 2015).

Hughes, R. (17–18 March, 2009) Engineering Process for IEC 61850, SEAPAC 2009. *CIGRE Australia Panel B5*, Melbourne.

IEC 61850–6, Communication Networks and Systems for Power Utility Automation – Part 6: Configuration Description Language for Communication in Electrical Substations Related to IEDs.

Kim, Y-H., Han, J.Y., Lee, Y.J., et al. (2011) Development of IEC 61850 based substation engineering tools with IEC 61850 schema library, *Smart Grid and Renewable Energy Review* 2 (Korea), 271–277.

Nibbio, N., Genier, M., Brunner, C., et al. (2010) Engineering Approach for the End User in IEC 61850 Applications, paper D2/B5–115, *CIGRE session*.

Rim, S-J., Zeng, S-W. and Lee, S-J. (2009) Development of an intelligent station HMI in IEC 61850 based substation, *Journal of Electrical & Technology* 4(1), 13–18.

Yang, C-W., Vyatkin, V., Nair, N-K.C. and Chouinard, J. (November 7–10, 2011) Programmable Logic for IEC 61850 Logical Nodes by means of IEC 61499, *IEEE 37th International Conference of Industrial Electronics Society (IECON 11)*, Melbourne.

19

Future Technological Trends

Two decades have now passed since the incursion of digital technology; nowadays, the power system infrastructure is impregnated on a significant scale with software-based solutions and communication networks. The Standard IEC 61850 dedicated to substation secondary systems (SAS) had been already adopted partially by utilities and other substation owners world-wide, although it is perceived as a complex set of documents. Even so, based on current facts, future trends may be clearly forecasted by a simple sentence: "intensive application of Standard IEC 61850". In that sense, the following four branches are visualized as priority targets:

1. Substation facilities and functionalities
2. Testing scope and procedures
3. Wide area control and monitoring
4. Smart grid solutions.

The scope explanation of these new expected applications is the purpose of this chapter.

19.1 Toward the Full Digital Substation

Although almost all devices applied now in SASs are based on digital technology, many utilities appear still to be reluctant to implement certain functionalities and facilities by using already well-developed digital techniques. An analysis of the main relegated items is presented next.

19.1.1 Horizontal Communication as per IEC 61850 (GOOSE Messaging)

Bay controllers perform part of their control and monitoring functions, such as Synchrocheck and switchgear status supervision, via direct interaction with the HV apparatus, but they need

Substation Automation Systems: Design and Implementation, First Edition. Evelio Padilla.

to share time-critical information in order to be able to carry out other essential functions such as control command delivery. The reason comes from the safety restriction that the circuit breaker and disconnector must be operated only when other associated switchgear is/are in position (open or closed), so that no failure will occur due the switching operation. That information sharing is given by the inter-bay interlocking logic mandatorily implemented in SAS projects.

The reluctance for use comes from a lack of confidence in the communication infrastructure to ensure the right data transference; preferring hard-wired connections for the communication between bay controllers instead of accepting the application of the GOOSE service. It is expected that confidence will be gained by an in-depth understanding of the robustness of communication networks and particularly the effectiveness of their self-monitoring attributes.

19.1.2 Unconventional Instrument Transformers

Utilities were afraid to use digital instrument transformers, mainly because of the application of proprietary communication protocols into original designs of that equipment. Such feelings will disappear with the emergence of the new generation of optical instrument transformers, which are IEC 61850 compliant.

19.1.3 Process Bus as Defined by IEC 61850–9-2

Although the benefits of the process bus concept defined by the Standard IEC 61850–9-2 are quite clear, until now only a few new substation projects around the world have been equipped with these systems. Some utilities have started doing pilot projects. Others are putting such systems in parallel with another type of solution for reliability.

Considering that no key factor to trip up the use of this solution is on the horizon, a massive application is expected medium/long term.

19.2 Looking for New Testing Strategies on SAS Schemes

Existing technical standards provide exhaustive guidelines for performing type and conformance tests on IEDs. However, none of them cover the testing of complete digital functional chain including the use of station bus for the transmission of control commands and other signals, and the process bus for the transmission of current and voltage values coming from HV instrument transformers.

The testing of these full digital functional chains has become an important issue with the deployment of the IEC 61850-based substations using standalone merging units, switchgear drivers and non-conventional instrument transformers, and featuring both the station bus (IEC 61850–8-1) and process bus (IEC 61850–9-2). The use of digital buses for the transmission of sampled values and signals adds new test cases to the current functional tests for SASs, including protection schemes. Also, the design of these systems has to take into account the constraints related to functional tests after the commissioning of the system, for example in the maintenance stage.

Because of this, a CIGRE Working Group (WG B5.53) is researching on the subject of new testing strategies for SASs covering the complete digital functional chain. The intended scope of the work includes:

- Qualification procedures for standalone merging units, non-conventional instrument transformers and serial communication links.
- Interoperability tests.
- Factory Acceptance Tests (FAT).
- Commissioning and Site Acceptance Tests.
- Maintenance tests after commissioning.

The work also intends to deliver recommendations related to type tests applicable on SAS solutions, such as:

- Testing of complete functional chains under different possible topologies.
- Test of functions acquiring values from different sources.
- Test scenarios related to communication troubles like loss of sampled values, loss of time synchronization signal.
- Functional interoperability tests, including in particular cases the effect of transients signals of different frequencies.
- Isolation means of test objects for FATs, Site Acceptance Tests and Maintenance Tests.
- Possible approaches for partial testing of a function with verification of recovery of the different units tests.
- Accuracy tests.

The result of the mentioned work will be published through CIGRE media as a paper in *Electra Review*, a tutorial or a Technical Brochure.

19.3 Wide Area Control and Monitoring Based on the IEC/TR 61850–90–5

From the analysis of several events recently occurring in power systems, the need for appropriated resources for transmitting states information and time-synchronized power measurements over wide area networks in order to enable centralized evaluations and decision-capabilities on remote locations was identified. This led to development of the Synchrophasor technology to which the IEC/TR 61850–90–5 is dedicated. The new technology monitors electrical waves in the power system and synchronizes those measurements to a common time source. It characterizes system conditions by calculating phasors (amplitude and phase) at a number of samples per second. To accomplish that functionality, two new devices are now connected to the station bus of the substation: Phasor Measurement Units (PMU) and Phasor Data Concentrators (PDC). The PMU measure or calculate frequency, voltage, current and phase angles, time-synchronized across the entire power system. These measurements, taken from diverse points across the system, allow unprecedented visibility and better situational awareness of real-time power system condition. Being a very useful technology, it is expected that it will be widely applied in forthcoming years.

19.4 Integration of IEC 61850 Principles into Innovative Smart Grid Solutions

Smart grids are electrical systems in which information and communication technologies have a preponderant role. One of the relevant characteristics of smart grids is the ability of their devices to communicate and exchange data with each other. Already recognized the Standard IEC 61850 as the open standard that enables interoperability between devices, its use has recently begun and will continue in future penetration of the smart grid domain.

Further Reading

Apostolov, A. (April 25–26. 2005) IEC 61850 and Disturbance Recording, Georgia Tech Fault and Disturbance Analysis Conference, Atlanta.

Cardenas, J., De Viñaspre, A.L., Argandoña, R., et al. (2012) The next generation of smart substations. challenges and possibilities, Paper B5–110. *CIGRE session.*

Dolezilek, D., Whitehead, D. and Skendzic, V. (2010) Integration of IEC 61850 GSE and Sampled Value to Reduce Substation Wiring, Paper for Schweitzer Engineering Laboratories, Inc.

Eger, K., Rusitschka, S. and Gerdes, C. (6–9, June 2011) A Plug and Play Concept for IEC 61850 in a Smart Grid, *21st International Conference on Electricity Distribution CIRED*, Frankfurt.

Falk, H. (December 2012) IEC 61850–90–5 an Overview, *PAC Review*, 39–43.

Filho, J.M.O., Sollero, R.B., Bass, W., et al. (2010) Substation automation in the next decade: Predictable steps and sound visions, Paper B5–110. *CIGRE session.*

IEEE C37.118.1, Standard for Synchrophasor Measurements for Power System.

Lemos, M., Santos, J.M., Varela, G., et al. (6–9 June 2011) Substation Automation Systems current challenges and future requirements – The impact Project perspective, Paper 0504, *21st International Conference on Electricity Distribution CIRED*, Frankfurt.

Richards, S., Pavaiya, N., Boucherit, M., et al. (June 2014) Digital substations: Feedback on site experience, *PAC World.*

Ridwan, M.I., Miswan, N.S., Noran, M.N., et al. (2014) Testing of IEC 61850 Compliant Smart Grid Devices: A Malaysian Experience, Paper C4–1093, *CIGRE - AORC Technical Meeting.*

Skendzic, V., Ender, I. and Zweigle, G. (2007) IEC 61850–9-2 Process Bus and Its impact on Power System Protection and Control Reliability, Paper for Schweitzer Engineering laboratories, Inc.

Wu, F.F., Moslehi, K. and Bose, A. (November 2005) Power system control centers: Past, present and future, *Proceedings of the IEEE*, 93(11).

Zhabelova, G., Yang, C.W. and Vyatkin, V. (July 29–31. 2013) SysGrid: IEC 61850/IEC 61499 Based Engineering Process for smart Grid Automation Design, *IEEE Conference on Industrial Informatics (INDIN 13)*, Bochum.

Appendix A

Samples of Equipment and System Signal Lists

A.1 Signals List Related to Circuit Breakers (Each One)

Description	Type	Status
Position indication pole A	Input	Open
Position indication pole A	Input	Closed
Position indication pole B	Input	Open
Position indication pole B	Input	Closed
Position indication pole C	Input	Open
Position indication pole C	Input	Closed
Local/remote selector	Input	Local
Local/remote selector	Input	Remote
Gas pressure low	Input	Normal
Gas pressure low	Input	Block
Oil pressure low	Input	Normal
Oil pressure low	Input	Block
Trip circuit I failure	Input	Normal
Trip circuit I failure	Input	Alarm
Trip circuit II failure	Input	Normal
Trip circuit II failure	Input	Alarm
Oil pump faulty	Input	Normal
Oil pump faulty	Input	Block
MCB hydraulic pump open	Input	Normal
MCB hydraulic pump open	Input	Alarm
Oil pump overload	Input	Normal
Oil pump overload	Input	Alarm
Oil pump overrun	Input	Normal

(*continued*)

Substation Automation Systems: Design and Implementation, First Edition. Evelio Padilla.

(*continued*)

Description	Type	Status
Oil pump overrun	Input	Alarm
DC voltage lost	Input	Normal
DC voltage lost	Input	Alarm
AC voltage lost	Input	Normal
AC voltage lost	Input	Alarm
Pole discrepancy condition	Input	Normal
Pole discrepancy condition	Input	Alarm
Pole discrepancy trip	Input	Normal
Pole discrepancy trip	Input	Trip
Opening command circuit I	Output	Open
Opening command circuit II	Output	Open
Closing command	Output	Close

A.2 Signals List Related to Collateral Devices

Description	Type	Status
Busbar/Main protection	Input	Normal
Busbar/Main protection	Input	Failure
Busbar/BF protection	Input	Normal
Busbar/BF protection	Input	Failure
Trafo/Main protection	Input	Normal
Trafo/Main protection	Input	Failure
Trafo/Secondary protection	Input	Normal
Trafo/Secondary protection	Input	Failure
Line 1/Main protection	Input	Normal
Line 1/Main protection	Input	Failure
Line 1/Backup protection	Input	Normal
Line 1/Backup protection	Input	Failure
Line 2/Main protection	Input	Normal
Line 2/Main protection	Input	Failure
Line 2/Backup protection	Input	Normal
Line 2/Backup protection	Input	Failure
Line n/Main protection	Input	Normal
Line n/Main protection	Input	Failure
Line n/Backup protection	Input	Normal
Line n/Backup protection	Input	Failure
Autoreclosure relay	Input	Normal
Autoreclosure relay	Input	Failure

A.3 Signals List Related to the Auxiliary Power System

Description	Type	Status
Position indication CB A	Input	Open
Position indication CB A	Input	Closed
Position indication CB B	Input	Open
Position indication CB B	Input	Closed
Position indication CB C	Input	Open
Position indication CB C	Input	Closed
Local/Remote selector CB A	Input	Local
Local/Remote selector CB A	Input	Remote
Local/Remote selector CB B	Input	Local
Local/Remote selector CB B	Input	Remote
Local/Remote selector CB C	Input	Local
Local/Remote selector CB C	Input	Remote
Spring condition CB A	Input	Normal
Spring condition CB A	Input	Discharged
Spring condition CB B	Input	Normal
Spring condition CB B	Input	Discharged
Spring condition CB C	Input	Normal
Spring condition CB C	Input	Discharged
Trip circuit failure CB A	Input	Normal
Trip circuit failure CB A	Input	Alarm
Trip circuit failure CB B	Input	Normal
Trip circuit failure CB B	Input	Alarm
Trip circuit failure CB C	Input	Normal
Trip circuit failure CB C	Input	Alarm
MCB motor open CB A	Input	Normal
MCB motor open CB A	Input	Alarm
MCB motor open CB A	Input	Normal
MCB motor open CB A	Input	Alarm
MCB motor open CB A	Input	Normal
MCB motor open CB A	Input	Alarm
DC voltage lost CB A	Input	Normal
DC voltage lost CB A	Input	Alarm
DC voltage lost CB B	Input	Normal
DC voltage lost CB B	Input	Alarm
DC voltage lost CB C	Input	Normal
DC voltage lost CB C	Input	Alarm
AC voltage lost CB A	Input	Normal
AC voltage lost CB A	Input	Alarm
AC voltage lost CB B	Input	Normal
AC voltage lost CB B	Input	Alarm
AC voltage lost CB C	Input	Normal
AC voltage lost CB C	Input	Alarm
Oil high temperature APT A	Input	Normal
Oil high temperature APT A	Input	Alarm
Oil high temperature APT B	Input	Normal
Oil high temperature APT B	Input	Alarm

(*continued*)

(*continued*)

Description	Type	Status
Oil low level APT A	Input	Normal
Oil low level APT A	Input	Alarm
Oil low level APT B	Input	Normal
Oil low level APT B	Input	Alarm
Opening command CB A	Output	Open
Closing command CB A	Output	Close
Opening command CB B	Output	Open
Closing command CB B	Output	Close
Opening command CB C	Output	Open
Closing command CB C	Output	Close

A.4 Signals List Related to the SAS Itself

Description	Type	Status
Feeder 1/Control Unit/Local remote selector	Input	Local
Feeder 1/Control Unit/Local remote selector	Input	Remote
Feeder 2/Control Unit/Local remote selector	Input	Local
Feeder 2/Control Unit/Local remote selector	Input	Remote
Feeder n/Control Unit/Local remote selector	Input	Local
Feeder n/Control Unit/Local remote selector	Input	Remote
Feeder 1/Control Unit condition	Input	Active
Feeder 1/Control Unit condition	Input	Failure
Feeder 2/Control Unit condition	Input	Active
Feeder 2/Control Unit condition	Input	Failure
Feeder n/Control Unit condition	Input	Active
Feeder n/Control Unit condition	Input	Failure
Communication network	Input	Normal
Communication network	Input	Failure
Trafo 1 Emergence Trip	Input	Normal
Trafo 1 Emergence Trip	Input	Trip
Trafo 2 Emergence Trip	Input	Normal
Trafo 2 Emergence Trip	Input	Trip

Appendix B

Project Drawing List: Titles and Contents

B.1 General Interest Drawings

Project drawing list:

- Drawing names
- Drawing numbers
- Date of issue and version
- Size of drawing format (A0, A3, ...)
- Number of pages

Substation general electromechanical layout:

- List of components
- List of symbols
- Location of: buildings, high voltage equipment, line entrances, transition towers, cabling channels, ducts, internal roads and protective fence
- Access road
- Geography coordinates
- List of reference drawings

Substations interconnection diagram

- Substation names and locations
- Voltage levels
- HV busbar arrangements
- Length of transmission lines
- Connection points
- Phase order
- List of reference drawings

Substation Automation Systems: Design and Implementation, First Edition. Evelio Padilla.

Substation/line transition tower (each tower)
- Tower profile
- Phase order
- Shielding conductor (substation side)
- Shielding conductor (line side)
- Shielding connection clamp
- Shielding conductor hardware
- OPGW cable
- Isolator strings
- Power conductor (substation side)
- Power conductor (line side)
- Power connection clamps
- Power conductor hardware
- Table with indication of responsible of supply and erection for each item (substation contractor or line contractor)
- List of reference drawings

Line entrances layout:
- Feeder names
- Phase order
- Distances
- Restrictive objects
- Entrance interfaces (e.g., undergrounding cables)
- Transition points between OPGW and other types of fiber-optic cable
- List of reference drawings

Earthing system:
- Buildings and equipment layout
- Earthing mesh layout
- Cable data
- Installation depth
- Dimensions
- Connections and details
- List of reference drawings

Works at remote locations (associated substations):
- Scope and details
- Intervention plans
- List of reference drawings

Spare parts – Master list:
- Item description
- Item codes
- Quantities
- Associated subsystem
- List of reference drawings

B.2 Electromechanical Drawings (High Voltage Equipment and Control Facilities)

High voltage cable list – Description and details:
- Item description
- Application reference
- List of reference drawings

Power Transformer – Dimension and specs:
- General view
- Location and details of accessories (current transformers, temperature monitors, valves, …)
- Cooler system outline
- Control box
- Interface cabinet
- MV cable connection box
- On-load tap changer control box
- Buchholz relay
- Bushing potential device
- List of reference drawings

Power Transformer – Schematic diagram:
- Table of graphic symbols
- Device lists
- Main and backup cooler circuits
- Alarm and trip circuits
- Bushing potential device circuit
- Fault indicator circuits
- Voltage regulating circuit
- List of reference drawings

Power transformer – Interface cabinet:
- General view
- Terminal layout
- Terminal identification (including spare terminals)
- Entrance of power distribution circuits
- List of reference drawings

Circuit breaker – Dimension and specs:
- General view
- Control box
- Earthing terminal
- Details of operating mechanism
- List of reference drawings

Circuit breaker – Schematic diagram:
- Table of graphic symbols
- Terminal identifications
- Trip circuits
- Closing circuit

- Trip circuits supervision
- AC/DC power circuits
- Auxiliary contacts (for position indication)
- Illumination circuits
- Heating circuit
- Alarm/blocking circuits
- List of reference drawings

Disconnector – Dimensions and specs (each type):
- General view
- Control box
- Earthing terminal
- Details of operating mechanism
- List of reference drawings

Disconnector – Schematic diagram:
- Table of graphic symbols
- Terminal identifications
- Opening circuit
- Closing circuit
- AC/DC power circuit
- Alarm/blocking circuits
- List of reference drawings

Current transformers – Dimensions and specs:
- General view
- Secondary box
- Fixing means
- Earthing terminal
- List of reference drawings

Current transformer – Secondary box:
- Terminal layout and identification
- List of reference drawings

Voltage transformer – Dimension and specs:
- General view
- Secondary box
- Earthing terminal
- Fixing means
- List of reference drawings

Voltage transformer – Secondary box:
- Terminal layout and identification
- List of reference drawings

Surge arrester – Dimension and specs:
- General view
- Secondary terminal
- Operation counter
- List of reference drawings

Surge arrester – Connection diagram:
- Earthing circuit
- List of reference drawings

Pothead – Dimension and specs:
- General view
- List of reference drawings

Pothead – Installation procedure:
- General description
- Earthing connection
- List of reference drawings

Three winding arrangement of power transformers – Details:
- General view
- LV circuits
- List of reference drawings

Lightning poles, loudspeaker and telephone – Layout and details:
- General description
- List of components
- Schematic diagrams
- List of reference drawings

Main Control Room MCR – Internal earthing grid:
- General layout
- Details of accessories and connections
- List of reference drawings

MCR – Electromagnetic shielding:
- General description
- Calculations
- List of reference drawings

Local Control Room LCR – Internal earthing grid:
- General layout
- Details of accessories and connections
- List of reference drawings

LCR – Electromagnetic shielding:
- General description
- Calculations
- List of reference drawings

B.3 Electromechanical Drawings (Control, Protection, Measurement and Communications)

Single-line diagram (including future):
- Table of graphic symbols
- Bay/feeder names/numbers
- Interconnections (primary power circuits)
- Codes of HV equipment
- Details of relevant parts like power transformer
- List of reference drawings

Three-line diagram:
- Table of graphic symbols
- Table of relay and other IED identification

- Bay/feeder names/numbers
- Interconnections (primary power circuits)
- Codes of HV equipment
- Links between IEDs and instrument transformers
- List of reference drawings

External alarm annunciation:
- General description
- Interface with other IEDs
- List of selected alarms
- Alarm grouping criteria
- List of reference drawings

Control and signalizing principles:
- Table of device identification
- Table of symbols identification
- Cubicle identification
- Function identification
- Device locations
- Control and signalizing circuits
- Control structure at different levels (linking diagram)
- List of reference drawings

Description of signals and conventions:
- Code structure
- Applicable standard
- Examples
- List of reference drawings

Electrical interlocking principles:
- Single-line diagram (partial displays)
- Table of symbols identification
- Interlocking equations
- List of reference drawings

Electrical interlocking logic diagram:
- Single-line diagrams
- Lists of symbols and abbreviations
- Interlocking equations
- List of reference drawings

Voltage regulation and parallel operation – Principles:
- HV Equipment for reference voltage
- Location of regulating devices
- Control circuit
- List of reference drawings

Synchronization – Principles:
- Table of graphic symbols
- List of abbreviations
- Cubicle identification
- HV equipment for reference voltage
- Indication of secondary windings to be used

- Voltage circuits
- Contacts chains
- Terminals identifications
- List of reference drawings

SAS topology:
- Table of device identification
- Block diagram
- HMI
- Links between devices (LAN, station bus, process bus)
- Details of connection points
- Devices locations
- Networking devices
- Collateral IEDs (such as protective relays and external alarm annunciator)
- Time signal circuit
- Details of communication path to the remote control center
- List of reference drawings

Fiber-optic network – Principle:
- Table of patch cord types
- Length on patch cords
- Device identification and location
- Cubicle identification
- Terminals identification
- Connection points
- List of reference drawings

Control signal list:
- Bay/feeder identification
- Object identification
- Signal text at bay controller
- Signal text at HMI
- Signal text at NCC

Control – parameters for signals configuration

Control software list

Control IEDs – Lists and data sheets:
- List of devices and application
- Technical data
- List of reference drawings

Control IEDs – Technical catalog compendium

Control, protection and other cubicle identification:
- Code structure
- Applicable standard
- Examples
- Complete list of cubicles
- List of reference drawings

Control cubicle – View and details (each board):
- Dimensions
- Front view

- Device layout
- Identification plate
- Bay/feeder name
- Earthing terminal
- List of reference drawings

Control cubicle – Concept:
- Structure
- Materials
- Surface treatment
- Color
- Degree of protection (again environmental factors)
- Earthquake resistance
- Immunity again electromagnetic effects
- List of reference drawings

Control cubicle – Schematic diagram (each one):
- AC distribution circuits
- DC polarities distribution
- Switchgear control circuits
- Circuit breaker trip circuits
- Circuit breaker reclosing scheme
- Supervision of trip circuits
- Tap changer control circuit
- Interlocking schemes
- Position indication circuits
- Alarms and blocking circuits

Control cubicle – Erection

Human Machine Interface (HMI) – Description and details:
- System overview
- Components identifications
- Basic functions
- Colors and symbols
- System startup
- Alarm list
- Event list
- Pictures
- Maintenance and backup procedures
- List of reference drawings

HMI – Database (process database):
- Bay/feeder identification
- Object identifications
- Signal text
- State indication
- List of reference drawings

Bay Controller cubicle – Functional description:
- General system overview
- List of abbreviations

- List of symbols
- Introduction
- Process diagram
- Devices layout
- Hardware description
- Device compositions
- Functions descriptions
- Buttons identifications
- LEDs applications (assigned alarm signal)
- Self-supervision features
- Front view of the IED
- Protection interface
- Interface with upstream control level
- Power supply and distribution
- Programming tools
- List of reference drawings

Bay controller – Schematic diagram
Bay controller – Signal list:

- Single-line diagram of the associated bay/feeder
- BIO cards identifications
- Slots identifications
- Object identifications (HV apparatus, signal, function)
- Signals texts
- Signals types
- Signals positions
- Specific reference on schematic diagram
- List of reference drawings

Bay controller – Schematic diagram (software):

- BIO cards identifications
- Slots identifications
- Functions identifications
- Connections
- List of reference drawings

Bay controller Database
Station Controller – Functional description
Station Controller – Signal list
Auxiliary System Controller – Functional description:

- General system overview
- List of abbreviations
- List of symbols
- Introduction
- Process diagram
- Devices layout
- Hardware description
- Device compositions
- Functions descriptions

- Buttons identifications
- LEDs applications (assigned alarm signal)
- Self-supervision features
- Front view of the IED
- Interface with upstream control level
- Power supply and distribution
- Programming tools
- List of reference drawings

Auxiliary System Controller – Schematic diagram:

- BIO cards identifications
- Slots identifications
- Functions identifications
- Connections
- List of reference drawings

Auxiliary System Controller – Signal list
Auxiliary System Controller – Automatic transfer: Principles
Auxiliary System Controller – Point to point signal list
Control internal cabling (each cubicle):

- Cubicle identification
- Terminal identifications
- Cable description
- List of reference drawings

Control external cabling (each bay/feeder):

- Cubicle identifications
- Terminal identifications
- Cable description
- List of reference drawings

Control cable list
Control cable specifications:

- General description
- Technical data
- Screen characteristics
- List of reference drawings

Interface between disciplines – Principles
Signal distribution cubicles – Layout and details (each one):

- Internal view
- List of components
- Terminal blocks layout
- Terminal identifications
- List of reference drawings

Junction box – Details (each one):

- Internal view
- List of components
- Terminal blocks layout
- Terminal identifications
- List of reference drawings

Signal distribution cubicles – Internal cabling:
- Cubicle identification
- Terminals identification
- Internal electric bridges
- Cables identifications
- List of reference drawings

Protocol conversion
Schematic diagram (each bay and bar)
Fiber-optic cables – Identification and details:
- Cable identifications
- Patch cords details
- Accessories
- Associated subsystem
- List of reference drawings

Protection System – Principles
Protection – Schematic diagram (each type)
Protection and teleprotection relays and devices – Lists and data sheets
Protection cubicle concept
Protection software list
Protection relays and devices – Catalog compendium
Protection relay – Setting parameters
Internal protection cabling (each board)
External protection cabling
Protection cables list
Protection cubicles – View and details (each board)
Teleprotection – Principles:
- List of protection schemes
- Cubicle identifications
- Substation names
- Channel identifications
- Transmission media
- Cubicle location
- List of reference drawings

Teleprotection – schematic diagram
Teleprotection board – View and details
Signal distribution cubicles – View and details (each board)
Control external cabling – Principles:
- Set of complete circuits from HV apparatus to bay controllers
- Chain of cubicles including control boxes of HV apparatus, junction boxes and signal distribution cubicles
- Connections into each cubicle
- Terminal identifications
- Equipment/pole identifications
- Function of contacts
- Device identifications
- List of reference drawings

External connections cubicle (grouping) – View and details (each cubicle and type)
External connection cubicle – Cabling (each board)
Communication function – Principles
Communication cubicles – View and details (each cubicle)
Communication components – Lists and data sheets
Communication software list
Communication with NCC
Fault recording system
Energy metering cubicle – View and details
Other LV components – Lists and data sheets
Other components – Catalog compendium
Measurement function – Principles:

- Table of graphic symbols
- Secondary windings used for measurement purpose
- Location of measurement devices
- Communication interfaces
- List of reference drawings

Energy metering – Principle of communications:

- Device identification and description
- Locations identifications
- Connections diagram
- Communication media
- Details of communication interfaces
- List of reference drawings

Cabling principle
Cable – Description and details
Cabling routes
Channel and duct capabilities calculation
Fiber-optic network
Fiber-optic – Cable and connector list
Fiber-optic – Transition box
Fiber-optic – Distribution cubicle
Energy metering – Interconnection
Mechanical interlocking system – Principles:

- View of key devices
- Key identification
- Key functions
- Interlocking diagrams
- List of reference drawings

Main Control House (MCH) – Cubicles and other components layout:

- List of components identification
- Components location
- Dimensions
- Opening direction of cubicle doors
- List of reference drawings

MCH – Front view of cubicles

MCH – Control desk
MCH – internal cabling
Local Control Room (LCR) – Cubicles, transformers … layout:

- List of components identification
- Components location
- Location of wall type cabinets
- Dimensions
- Opening direction of cubicle doors
- List of reference drawings

LCR – Internal earth system:

- Front view of the building
- List of materials
- Cable layout
- Connection details
- Feeders for connection to the main earthing system
- List of reference drawings

LCR – Front view of cubicles

- Table of cubicles for identification
- Details of cubicle front views
- List of reference drawings

LCR – Cabling between cubicles
LCR – Electromagnetic shielding

- Mesh details
- Installation details
- Earth connections
- List of reference drawings

FAT program

B.4 Electromechanical Drawings (Auxiliary Power System)

Single-line diagram:

- Table of graphic symbols
- Feeder names/numbers
- Circuits arrangements
- Codes of MV and LV equipment
- Details of relevant parts like distribution transformer
- List of reference drawings

MV switchgear – Description and detail
MV switchgear – Schematic diagram:

- General system overview
- List of abbreviations
- List of symbols
- Process diagram
- Power circuits
- Control circuits

- Devices description and layout
- Buttons identifications
- Self-supervision features
- Interface with upstream control level
- Power supply and distribution
- List of reference drawings

Auxiliary System Room (ASR) – Components layout:

- List of components identification
- Components location
- Location of wall type cabinets
- Dimensions
- Opening direction of cubicle doors
- List of reference drawings

ASR – Internal earthing system:

- Front view of the building
- List of materials
- Cable layout
- Connection details
- Feeders for connection to the main earthing system
- List of reference drawings

Load study AC:

- Introduction
- Calculation procedure
- Use factors and criteria
- Calculation
- Results
- List of reference drawings

Load study DC:

- Introduction
- Calculation procedure
- Use factors and criteria
- Calculation
- Results
- List of reference drawings

AC Distribution – Principles (each voltage level):

- Cubicle identification
- Cubicle location
- AC circuits
- Interconnections
- Loads identifications
- List of reference drawings

DC Distribution – Principles:

- Cubicle identification
- Cubicle location
- AC circuits
- Interconnections

- Loads identifications
- List of reference drawings

Discrimination studies – AC:

- Introduction
- Concepts
- Single-line diagram
- Discrimination cases
- CB data and performance curves
- Results
- List of reference drawings

MCH electrical installations:

- Introduction
- Standards and criteria
- Power distribution scheme (AC and DC)
- Calculations
- Analysis of illumination level
- Power plugs
- Cubicle loads
- Air conditioned effect
- Illumination devices characteristics and layout
- Load study
- Feeder calculations
- Lists and data sheets of components
- Conduits and accessories
- List of reference drawings

LCR electrical installations:

- Introduction
- Standards and criteria
- Power distribution scheme (AC and DC)
- Calculations
- Analysis of illumination level
- Power plugs
- Cubicle loads
- Air conditioned effect
- Illumination devices characteristics and layout
- Load study
- Feeder calculations
- Lists and data sheets of components
- Conduits and accessories
- List of reference drawings

Auxiliary system – Schematic diagram
Auxiliary system local and remote operation – Principles and details
Auxiliary System – Software list
DC voltage distribution cubicle – View and details (each cubicle):

- Front view
- Devices identification main technical data and amounts

- Cubicle dimensions
- Device layout
- Opening direction of the door
- List of reference drawings

Auxiliary system cubicles – Internal cabling (each cubicle)
Auxiliary system cubicles – Erection
Distribution Centre A – Power and control principle:

- Single-line diagram
- List of components
- List of symbols
- Feeder identifications
- Spare circuits
- Cubicle identification
- Busbar identifications
- Measurement circuits
- List of reference drawings

Distribution Centre B – Power and control principles:

- Single-line diagram
- List of components
- List of symbols
- Feeder identifications
- Spare circuits
- Cubicle identification
- Busbar identifications
- Measurement circuits
- List of reference drawings

Automatic transfer system – Principle scheme (each one):

- Single-line diagram
- List of symbols
- Logic diagrams
- Timers
- List of reference drawings

Auxiliary system components – Lists and data sheets
Auxiliary voltages distribution – Cabling (each bay/feeder):

- Cubicle identification
- Terminal blocks identifications
- Terminal identifications
- Cables identifications and main technical data

LV Cables (AC) – List and description
LV Cables (AC) – Calculation:

- Introduction
- Distribution circuits
- Calculation procedure and criteria
- Calculations
- Results
- List of reference drawings

LV Cables (DC) – List and description

LV cables (DC) – calculation:

- Introduction
- Distribution circuits
- Calculation procedure and criteria
- Calculations
- Results
- List of reference drawings

Distribution transformer – Description and details:

- General view
- Dimensions
- List of devices and accessories
- Technical data
- Weight
- Fixing facilities
- Terminal boxes
- Earthing terminal
- Schematic diagram

Distribution transformers – Calculation

Distribution transformer – Erection

Distribution transformers – Drawing compendium

Earthing transformers – Layout and details

Earthing transformer – Calculation

Earthing transformer – Erection

Earthing transformer – Drawing compendium

Earthing resistance – Description and details:

- General view
- Dimensions
- List of devices and accessories
- Technical data
- Weight
- Fixing facilities
- Connection details
- Terminal boxes
- Earthing terminal
- Schematic diagram
- List of reference drawings

Earthing resistance – Calculation

Earthing resistance – Erection

Earthing resistance – Drawing compendium

MCH electrical installation – Layout and details

MCH electrical installation – Calculation

LCR electrical installation – Layout and details

LCR electrical installation calculation

Auxiliary System Room (ASR) – External cabling

ASR – Power distribution cabling

ASR – Cable and accessories list (AC and DC)
ASR – Cable calculation (AC and DC)
ASR – Components layout
ASR – Front view of cubicles
ASR – Internal cabling
ASR – Air extractor
Medium voltage cable – Shield earthing
Medium voltage cable – Calculation
Medium voltage cubicle – Layout, details and connections
Medium voltage cubicle – Schematic diagram
Medium voltage cubicle – Cabling
Medium voltage cubicle – Drawing compendium
Batteries – Layout, details and connections:

- General views
- List and data of components
- Components layout
- Connections
- Fixing detail
- Earthing details
- Monitoring circuits
- List of reference drawings

Batteries and chargers – Calculation (each voltage level):

- Introduction
- Reference documents
- Environmental conditions
- Design criteria
- Calculation under different load conditions
- Results
- List of reference drawings

Fuse box – Layout, details and connections:

- Front and sides views
- Technical and general data
- Dimensions
- Earthing terminal
- Holes diameters
- Fixing facilities
- List of reference drawings

Batteries charger – Layout, details and connections:

- Schematic diagram
- Three-line diagram
- List of symbols and abbreviations
- Front and profiles view
- Mounting facilities
- Fixing facilities
- Components layout
- List of reference drawings

Battery charger – Calculation
Lightning system – Layout and details
Lightning system – Calculation
Fire detection system – Layout and details
Fire detection system – Calculation
External lightning diagram
External power plug
Access road lightning system – Layout and details
Access road lightning system – Calculation
Diesel generator – Description and details:

- General view
- Dimensions
- List of devices and accessories
- Technical data
- Weight
- Fixing facilities
- Terminal boxes
- Earthing terminal
- Schematic diagram
- List of reference drawings

Diesel generator – Drawing compendium
Diesel generator – Exhaust system calculation
Timing distribution system layout – Details and connections
Telephone system – Layout, details and connections
Loudspeaker system – Layout, details and connections
Medium voltage feeder – Layout and details
Medium voltage feeder – Calculation:

- Introduction
- Cables, description, details and applications
- Cable length
- Description of calculation procedure
- Voltage downfall considerations
- Thermal considerations
- Installation considerations
- Calculations
- List of reference drawings

Appendix C

Essential Tips Related to Networking Technology

C.1 Computer Network

A computer network or data network is a telecommunications network that allows computers to exchange data. In computer networks, networked computing devices pass data to each other along data connections (network links). Data is transferred in the form of packets. The connections between nodes are established using either cable media or wireless media. The best-known computer network is the Internet.

Network computer devices that originate, route and terminate the data are called network nodes. Nodes can include hosts such as personal computers, servers as well as networking hardware. Two such devices are said to be networked together when one device is able to exchange information with the other device, whether or not they have a direct connection to each other.

Computer networks differ in the transmission media used to carry their signals, the communications protocols to organize network traffic, the network's size, topology and organizational intent. In most cases, communications protocols are layered on (i.e., work using) other more specific or more general communications protocols.

Computer networks support applications such as access to the World Wide Web, shared use of application and storage servers, printers and fax machines, and use of email and instant messaging applications.

A computer network, or simply a network, is a collection of computers and other hardware components interconnected by communication channels that allow sharing of resources and information. Today, computer networks are the core of modern communication. All modern aspects of the public switched telephone network (PSTN) are computer controlled. Telephony increasingly runs over the Internet Protocol, although not necessarily the public Internet.

Substation Automation Systems: Design and Implementation, First Edition. Evelio Padilla.

The scope of communication has increased significantly in the past decade. This boom in communications would not have been possible without the progressive advancement of the computer network. Computer networks, and the technologies that make communication between networked computers possible, continue to drive computer hardware, software and peripheral industries. The expansion of related industries is mirrored by growth in the numbers and types of people using networks, from the researcher to the home user.

Computer networking may be considered a branch of electrical engineering, telecommunications, computer science, information technology or computer engineering, since it relies upon the theoretical and practical application of the related disciplines.

A computer network facilitates interpersonal communications allowing people to communicate efficiently and easily via email, instant messaging, chat rooms, telephone, video telephone calls and video conferencing. Providing access to information on shared storage devices is an important feature of many networks. A network allows sharing of files, data and other types of information giving authorized users the ability to access information stored on other computers on the network. A network allows sharing of network and computing resources. Users may access and use resources provided by devices on the network, such as printing a document on a shared network printer.

C.1.1 Data

Data is a set of values of qualitative or quantitative variables; restated, pieces of data are individual pieces of information. Data is measured, collected and reported and analyzed, whereupon it can be visualized using graphs or images.

Data as an abstract concept can be viewed as the lowest level of abstraction, from which information and then knowledge is derived.

Raw data, that is, unprocessed data, refers to a collection of numbers and characters and is a relative term. Data processing commonly occurs in stages, and the "processed data" from one stage may be considered the "raw data" of the next. Field data refers to raw data that is collected in an uncontrolled *in situ* environment. Experimental data refers to data that is generated within the context of a scientific investigation by observation and recording.

C.1.1.1 Meaning of Data, Information and Knowledge

Data, information and knowledge are closely related terms, but each has its own role in relation to the other. Data is collected and analyzed to create information suitable for making decisions, while knowledge is derived from extensive amounts of experience dealing with information on a subject.

"Information" bears a diversity of meanings that ranges from the everyday to the technical. Generally speaking, the concept of information is closely related to notions of constraint, communication, control, data, form, instruction, knowledge, meaning, mental stimulus, pattern, perception and representation.

In some cases, the concept of a symbol is used to distinguish between data and information; data is a series of symbols, while information occurs when the symbols are used to refer to something.

It is people and computers who collect data and impose patterns on it. These patterns are seen as information that can be used to enhance knowledge. These patterns can be interpreted as truth, and are authorized as aesthetic and ethical criteria. Events that leave behind perceivable physical or virtual remains can be traced back through data. Marks are no longer considered data once the link between the mark and observation is broken.

Computers use a binary alphabet, that is, an alphabet of two characters, typically denoted "0" and "1". More familiar representations, such as numbers or letters, are then constructed from the binary alphabet.

Some special forms of data are distinguished. A computer program is a collection of data, which can be interpreted as instructions. Most computer languages make a distinction between programs and the other data on which programs operate, but in some languages, notably Lisp and similar languages, programs are essentially indistinguishable from other data. It is also useful to distinguish metadata, that is, a description of other data. A similar yet earlier term for metadata is "ancillary data". The prototypical example of metadata is the library catalog, which is a description of the contents of books.

C.1.1.2 Data Modeling

Data modeling is a process used to define and analyze data requirements needed to support the business processes within the scope of corresponding information systems in organizations. Therefore, the process of data modeling involves professional data modelers working closely with business stakeholders, as well as potential users of the information system.

There are three different types of data models produced while progressing from requirements to the actual database to be used for the information system. The data requirements are initially recorded as a conceptual data model, which is essentially a set of technology independent specifications about the data and is used to discuss initial requirements with the business stakeholders. The conceptual model is then translated into a logical data model, which documents structures of the data that can be implemented in databases. Implementation of one conceptual data model may require multiple logical data models. The last step in data modeling is transforming the logical data model to a physical data model that organizes the data into tables, and accounts for access, performance and storage details. Data modeling defines not just data elements, but also their structures and the relationships between them.

Data modeling techniques and methodologies are used to model data in a standard, consistent, predictable manner in order to manage it as a resource. The use of data modeling standards is strongly recommended for all projects requiring a standard means of defining and analyzing data within an organization, for example, using data modeling:

- to assist business analysts, programmers, testers, manual writers, IT package selectors, engineers, managers, related organizations and clients to understand and use an agreed semi-formal model the concepts of the organization and how they relate to one another,
- to manage data as a resource,
- for the integration of information systems and
- for designing databases/data warehouses (aka data repositories).

Data modeling may be performed during various types of projects and in multiple phases of projects. Data models are progressive; there is no such thing as the final data model for a business or application. Instead a data model should be considered a living document that will change in response to a changing business. The data models should ideally be stored in a repository so that they can be retrieved, expanded and edited over time.

- *Strategic data modeling*: This is part of the creation of an information systems strategy, which defines an overall vision and architecture for information systems is defined. Information engineering is a methodology that embraces this approach.
- *Data modeling during systems analysis*: In systems analysis logical data models are created as part of the development of new databases.

Data modeling is also used as a technique for detailing business requirements for specific databases. It is sometimes called *database modeling* because a data model is eventually implemented in a database.

C.1.1.3 Data Type

In computer science and computer programming, a data type or simply type is a classification identifying one of various types of data, such as real, integer or Boolean, which determines the possible values for that type; the operations that can be done on values of that type; the meaning of the data and the way values of that type can be stored.

C.1.1.4 Network Packet

Computer communication links that do not support packets, such as traditional point-to-point telecommunication links, simply transmit data as a bit stream. However, most information in computer networks is carried in *packets*. A network packet is a formatted unit of data (a list of bits or bytes, usually a few tens of bytes to a few kilobytes long) carried by a packet-switched network.

In packet networks, the data is formatted into packets that are sent through the network to their destination. Once the packets arrive they are reassembled into their original message. With packets, the bandwidth of the transmission medium can be better shared among users than if the network were circuit switched. When one user is not sending packets, the link can be filled with packets from others users and so the cost can be shared with relatively little interference provided the link isn't overused.

Packets consist of two kinds of data: control information and user data (also known as payload). The control information provides data the network needs to deliver the user data, for example: source and destination network addresses, error detection codes and sequencing information. Typically, control information is found in packet headers and trailers, with payload data in between.

Often the route a packet needs to take through a network is not immediately available. In that case the packet is queued and waits until a link is free.

C.2 Network Topology

The physical layout of a network is usually less important than the topology that connects network nodes. Most diagrams that describe a physical network are therefore topological, rather than geographic. The symbols on these diagrams usually denote network links and network nodes.

C.2.1 Network Links

The transmission media (often referred to in the literature as the *physical media*) used to link devices to form a computer network include electrical cable, optical fibers (fiber-optic communication) and radio waves (wireless networking).

A widely adopted *family* of transmission media used in local area network (LAN) technology is collectively known as the Ethernet. The media and protocol standards that enable communication between networked devices over Ethernet are defined by IEEE 802.3. The Ethernet transmits data over both copper and optical fiber cables.

C.2.1.1 Wired Technologies

The orders of the following wired technologies are, roughly, from slowest to fastest transmission speed:

- *Twisted pair wire* is the most widely used medium for all telecommunication. Twisted-pair cabling consists of copper wires that are twisted into pairs. Ordinary telephone wires consist of two insulated copper wires twisted into pairs. Computer network cabling (wired Ethernet as defined by IEEE 802.3) consists of four pairs of copper cabling that can be utilized for both voice and data transmission.
- *Coaxial cable* is widely used for cable television systems, office buildings and other worksites for local area networks. The cables consist of copper or aluminum wire surrounded by an insulating layer, which itself is surrounded by a conductive layer. The insulation helps minimize interference and distortion.
- An *optical fiber* is a glass fiber. It carries pulses of light that represent data. Some advantages of optical fibers over metal wires are very low transmission loss and immunity from electrical interference. Optical fibers can be used for long runs of cable carrying very high data rates and are used for undersea cables to interconnect continents.

C.2.1.2 Wireless Technologies

Computers are very often connected to networks using the following wireless links:

- Terrestrial microwaves.
- Communications satellites.

C.2.2 Network Nodes

Apart from any physical transmission medium there may be, networks comprise additional basic system building blocks, such as a network interface controller (NICs), repeaters, hubs, bridges, switches, routers and modems.

C.2.3 Network Interface Controllers

A network interface controller (NIC) is computer hardware that provides a computer with the ability to access the transmission media and has the ability to process low-level network information. For example, the NIC may have a connector for accepting a cable, or an aerial for wireless transmission and reception, and the associated circuitry.

The NIC responds to traffic addressed to a network address for either the NIC or the computer as a whole.

C.2.4 Repeaters and Hubs

A repeater is an electronic device that receives a network signal, cleans it of unnecessary noise and regenerates it. The signal is retransmitted at a higher power level, or to the other side of an obstruction, so that the signal can cover longer distances without degradation.

A repeater with multiple ports is known as a hub. Repeaters require a small amount of time to regenerate the signal. This can cause a propagation delay that affects network performance. As a result, many network architectures limit the number of repeaters that can be used.

Hubs have been mostly obsolete by modern switches; but repeaters are used for long distance links, notably undersea cabling.

C.2.5 Bridges

A network bridge connects and filters traffic between two network segments to form a single network. This breaks the network's collision domain but maintains a unified broadcast domain. Network segmentation breaks down a large, congested network into an aggregation of smaller, more efficient networks.

C.2.6 Switches

A network switch is a device that forwards and filters datagrams between ports based on the MAC addresses in the packets. A switch is distinct from a hub in that it only forwards the frames to the physical ports involved in the communication rather than all ports connected. It can be thought of as a multi-port bridge.

C.2.7 Routers

A router is an internetworking device that forwards packets between networks by processing the routing information included in the packet or datagram. The routing information is often processed in conjunction with the routing table (or forwarding table). A router uses its routing table to determine where to forward packets.

C.2.8 Modems

Modems (MOdulator–DEModulator) are used to connect network nodes via wire not originally designed for digital network traffic or for wireless. To do this one or more frequencies are modulated by the digital signal to produce an analog signal that can be tailored to give the required properties for transmission.

C.3 Network Structure

Network topology is the layout or organizational hierarchy of interconnected nodes of a computer network. Different network topologies can affect throughput, but reliability is often more critical. With many technologies, such as bus networks, a single failure can cause the whole network to fail entirely. In general the more interconnections there are, the more robust the network is; but the more expensive it is to install.

C.3.1 Common Network Layouts

Common layouts are:

- *A bus network*: All nodes are connected to a common medium along this medium.
- *A star network*: All nodes are connected to a special central node.
- *A ring network*: Each node is connected to its left and right neighbor node, such that all nodes are connected and that each node can reach all other nodes by traversing nodes to the left or right.
- *A mesh network*: Each node is connected to an arbitrary number of neighbors in such a way that there is at least one traversal from any node to any other.
- *A fully connected network*: Each node is connected to every other node in the network.
- *A tree network*: Nodes are arranged hierarchically.

C.4 Communication Protocols

A communications protocol is a set of rules for exchanging information over network links.

Communication protocols have various characteristics. They may be connection-oriented or connectionless, they may use circuit mode or packet switching and they may use hierarchical addressing or flat addressing.

There are many communication protocols, a few of which are described next.

C.4.1 Ethernet

Ethernet is a family of computer networking technologies for local area networks (LANs) and metropolitan area networks (MANs). It was commercially introduced in 1980 and it was first standardized in 1983 as IEEE 802.3. It has since been refined to support higher bit rates and longer link distances. Over time, Ethernet has largely replaced competing wired LAN technologies such as token ring, FDDI and ARCNET.

Systems communicating over Ethernet divide a stream of data into shorter pieces called frames. Each frame contains source and destination addresses and error-checking data so that damaged data can be detected and re-transmitted.

Since its commercial release, Ethernet has retained a good degree of backward compatibility. Features such as the 48-bit MAC address and Ethernet frame format have influenced other networking protocols.

C.4.2 The Internet Protocol Suite

This is the computer networking model and set of communications protocols used on the Internet and similar computer networks. It is commonly known as TCP/IP, because its most important protocols, the Transmission Control Protocol (TCP) and the Internet Protocol (IP), were the first networking protocols defined in this standard.

TCP/IP provides end-to-end connectivity specifying how data should be packetized, addressed, transmitted, routed and received at the destination. This functionality is organized into four abstraction layers, which are used to sort all related protocols according to the scope of networking involved. From lowest to highest, the layers are: the link layer, containing communication technologies for a single network segment (link); the Internet layer, connecting hosts across independent networks, thus establishing internetworking; the transport layer handling host-to-host communication and the application layer, which provides process-to-process application data exchange.

The TCP/IP model and related protocol models are maintained by the Internet Engineering Task Force (IETF)

C.4.3 SONET/SDH

Synchronous Optical Networking (SONET) and Synchronous Digital Hierarchy (SDH) are standardized multiplexing protocols that transfer multiple digital bit streams over optical fiber using lasers. They were originally designed to transport circuit mode communications from a variety of different sources, primarily to support real-time, uncompressed, circuit-switched voice encoded in PCM (Pulse-Code Modulation) format. However, due to its protocol neutrality and transport-oriented features, SONET/SDH also was the obvious choice for transporting Asynchronous Transfer Mode (ATM) frames.

C.4.4 Asynchronous Transfer Mode

ATM is a switching technique for telecommunication networks. It uses asynchronous time-division multiplexing and encodes data into small, fixed-sized cells. This differs from other protocols such as the Internet Protocol Suite or Ethernet that use variable sized packets or frames. ATM has similarity to both circuit and packet switched networking. This makes it a good choice for a network that must handle both traditional high-throughput data traffic and real-time, low-latency content such as voice and video. ATM uses a connection-oriented model in which a virtual circuit must be established between two endpoints before the actual data exchange begins.

While the role of ATM is diminishing in favor of next-generation networks, it still plays a role in the last mile, which is the connection between an Internet service provider and the home user.

C.4.5 Basic Requirements of Protocols

Messages are sent and received on communicating systems to establish communications. Protocols should therefore specify rules governing the transmission. In general, much of the following should be addressed:

- *Data formats for data exchange.* Digital message bitstrings are exchanged. The bitstrings are divided in fields and each field carries information relevant to the protocol. Conceptually the bitstring is divided into two parts called the *header area* and the *data area*. The actual message is stored in the data area, so the header area contains the fields with more relevance to the protocol. Bitstrings longer than the maximum transmission unit (MTU) are divided in pieces of appropriate size.
- *Address formats for data exchange.* Addresses are used to identify both the sender and the intended receiver(s). The addresses are stored in the header area of the bitstrings, allowing the receivers to determine whether the bitstrings are intended for themselves and should be processed or should be ignored. A connection between a sender and a receiver can be identified using an address pair (*sender address*, *receiver address*). Usually some address values have special meanings. An all-*1*s address could be taken to mean an addressing of all stations on the network, so sending to this address would result in a broadcast on the local network. The rules describing the meanings of the address value are collectively called an *addressing scheme*.
- *Address mapping.* Sometimes protocols need to map addresses of one scheme on addresses of another scheme. For instance to translate a logical IP address specified by the application to an Ethernet hardware address. This is referred to as *address mapping*.
- *Routing.* When systems are not directly connected, intermediary systems along the *route* to the intended receiver(s) need to forward messages on behalf of the sender. On the Internet, the networks are connected using routers. This way of connecting networks is called *internetworking*.
- *Detection of transmission errors* is necessary on networks that cannot guarantee error-free operation. In a common approach, CRCs of the data area are added to the end of packets, making it possible for the receiver to detect differences caused by errors. The receiver rejects the packets on CRC differences and arranges somehow for retransmission.
- *Acknowledgement* of correct reception of packets is required for connection-oriented communication. Acknowledgements are sent from receivers back to their respective senders.
- *Loss of information: timeouts and retries.* Packets may be lost on the network or suffer from long delays. To cope with this, under some protocols, a sender may expect an acknowledgement of correct reception from the receiver within a certain amount of time. On timeouts, the sender must assume the packet was not received and retransmit it. In case of a permanently broken link, the retransmission has no effect so the number of retransmissions is limited. Exceeding the retry limit is considered an error.
- *Direction of information flow* needs to be addressed if transmissions can only occur in one direction at a time as on half-duplex links. This is known as Media Access Control. Arrangements have to be made to accommodate the case when two parties want to gain control at the same time.
- *Sequence control.* We have seen that long bitstrings are divided in pieces, and then sent on the network individually. The pieces may get lost or delayed or take different routes to their

destination on some types of networks. As a result pieces may arrive out of sequence. Retransmissions can result in duplicate pieces. By marking the pieces with sequence information at the sender, the receiver can determine what was lost or duplicated, ask for necessary retransmissions and reassemble the original message.
- *Flow control* is needed when the sender transmits faster than the receiver or intermediate network equipment can process the transmissions. Flow control can be implemented by messaging from receiver to sender.

Getting the data across a network is only part of the problem for a protocol. The data received has to be evaluated in the context of the progress of the conversation, so a protocol has to specify rules describing the context. These kinds of rules are said to express the *syntax* of the communications. Other rules determine whether the data is meaningful for the context in which the exchange takes place. These kinds of rules are said to express the *semantics* of the communications.

C.5 Geographical Scale of Network

A network can be characterized by its physical capacity or its organizational purpose. Use of the network, including user authorization and access rights, differ accordingly.

C.5.1 Local Area Network

A local area network (LAN) is a network that connects computers and devices in a limited geographical area such as a home, school, office building or closely positioned group of buildings. Each computer or device on the network is a node. Wired LANs are most likely based on Ethernet technology. Newer standards such as ITU-T G.hn also provide a way to create a wired LAN using existing wiring, such as coaxial cables, telephone lines and power lines.

The defining characteristics of a LAN, in contrast to a wide area network (WAN), include higher data transfer rates, limited geographic range and lack of reliance on leased lines to provide connectivity. Current Ethernet or other IEEE 802.3 LAN technologies operate at data transfer rates up to 10 Gbit/s.

C.5.2 Backbone Network

A backbone network is part of a computer network infrastructure that provides a path for the exchange of information between different LANs or sub-networks. A backbone can tie together diverse networks within the same building, across different buildings or over a wide area.

For example, a large company might implement a backbone network to connect departments that are located around the world. The equipment that ties together the departmental networks constitutes the network backbone. When designing a network backbone, network performance and network congestion are critical factors to take into account. Normally, the backbone network's capacity is greater than that of the individual networks connected to it.

Another example of a backbone network is the Internet backbone, which is the set of wide area networks (WANs) and core routers that tie together all networks connected to the Internet.

C.5.3 Wide Area Network

A wide area network (WAN) is a computer network that covers a large geographic area such as a city, country, or spans even intercontinental distances. A WAN uses a communications channel that combines many types of media such as telephone lines, cables and air waves. A WAN often makes use of transmission facilities provided by common carriers, such as telephone companies.

C.5.4 Intranet

An intranet is a set of networks that are under the control of a single administrative entity. The intranet uses the IP protocol and IP-based tools such as web browsers and file transfer applications. The administrative entity limits use of the intranet to its authorized users. Most commonly, an intranet is the internal LAN of an organization. A large intranet typically has at least one web server to provide users with organizational information. An intranet is also anything behind the router on a local area network.

C.5.5 Extranet

An extranet is a network that is also under the administrative control of a single organization, but supports a limited connection to a specific external network. For example, an organization may provide access to some aspects of its intranet to share data with its business partners or customers. These other entities are not necessarily trusted from a security standpoint. Network connection to an extranet is often, but not always, implemented via WAN technology.

C.6 Internetwork

An internetwork is the connection of multiple computer networks via a common routing technology using routers.

C.6.1 Internet

The Internet is the largest example of an internetwork. It is a global system of interconnected governmental, academic, corporate, public and private computer networks. It is based on the networking technologies of the Internet Protocol Suite. It is the successor of the Advanced Research Projects Agency Network (ARPANET) developed by DARPA of the United States Department of Defense. The Internet is also the communications backbone underlying the World Wide Web (WWW).

Participants in the Internet use a diverse array of methods of several hundred documented, and often standardized, protocols compatible with the Internet Protocol Suite and an addressing system (IP addresses) administered by the Internet Assigned Numbers Authority and address registries. Service providers and large enterprises exchange information about the reachability of their address spaces through the Border Gateway Protocol (BGP), forming a redundant worldwide mesh of transmission paths.

C.6.2 Routing

Routing is the process of selecting network paths to carry network traffic. Routing is performed for many kinds of networks, including circuit switching networks and packet switched networks.

In packet switched networks, routing directs packet forwarding (the transit of logically addressed network packets from their source toward their ultimate destination) through intermediate nodes. Intermediate nodes are typically network hardware devices such as routers, bridges, gateways or switches. General-purpose computers can also forward packets and perform routing, though they are not specialized hardware and may suffer from limited performance. The routing process usually directs forwarding on the basis of routing tables, which maintain a record of the routes to various network destinations. Thus, constructing routing tables, which are held in the router's memory, is very important for efficient routing. Most routing algorithms use only one network path at a time. Multipath routing techniques enable the use of multiple alternative paths.

There are usually multiple routes that can be taken, and to choose between them, different elements can be considered to decide which routes get installed into the routing table, such as (sorted by priority):

1. *Prefix-Length*: where longer subnet masks are preferred (independent if it is within a routing protocol or over different routing protocol).
2. *Metric*: where a lower metric/cost is preferred (only valid within one and the same routing protocol).
3. *Administrative distance*: where a lower distance is preferred (only valid between different routing protocols).

Routing, in a more narrow sense of the term, is often contrasted with bridging in its assumption that network addresses are structured and that similar addresses imply proximity within the network. Structured addresses allow a single routing table entry to represent the route to a group of devices. In large networks, structured addressing (routing, in the narrow sense) outperforms unstructured addressing (bridging). Routing has become the dominant form of addressing on the Internet. Bridging is still widely used within localized environments.

C.6.3 Network Service

Network services are applications hosted by servers on a computer network, to provide some functionality for members or users of the network, or to help the network itself to operate. The World Wide Web, email, printing and network file sharing are examples of well-known network services.

Services are usually based on a service protocol that defines the format and sequencing of messages between clients and servers of that network service.

C.6.4 Network Performance

C.6.4.1 Quality of Service

Depending on the installation requirements, network performance is usually measured by the quality of service of a telecommunications product.

There are many ways to measure the performance of a network, as each network is different in nature and design. Performance can also be modeled instead of measured. For example, state transition diagrams are often used to model queuing performance in a circuit-switched network. The network planner uses these diagrams to analyze how the network performs in each state, ensuring that the network is optimally designed.

C.6.4.2 Network Congestion

Network congestion occurs when a link or node is carrying so much data that its quality of service deteriorates. Typical effects include queuing delay, packet loss or the blocking of new connections. A consequence of these latter two is that incremental increases in offered load lead either only to small increase in network throughput, or to an actual reduction in network throughput.

Network protocols that use aggressive retransmissions to compensate for packet loss tend to keep systems in a state of network congestion – even after the initial load is reduced to a level that would not normally induce network congestion. Thus, networks using these protocols can exhibit two stable states under the same level of load. The stable state with low throughput is known as *congestive collapse.*

Modern networks use congestion control and congestion avoidance techniques to try to avoid congestion collapse.

C.6.4.3 Network Resilience

Network resilience is the ability to provide and maintain an acceptable level of service in the face of faults and challenges to normal operation.

C.6.5 Security Measures in Networks

C.6.5.1 Network Security

Network security consists of provisions and policies adopted by the network administrator to prevent and monitor unauthorized access, misuse, modification or denial of the computer network and its network-accessible resources. Network security is the authorization of access to data in a network, which is controlled by the network administrator. Users are assigned an ID and password that allows them access to information and programs within their authority.

Network security is used on a variety of computer networks, both public and private, to secure daily transactions and communications among businesses, government agencies and individuals.

C.6.5.2 Network Surveillance

Network surveillance is the monitoring of data being transferred over computer networks such as the Internet. The monitoring is often done surreptitiously and may be done by or at the behest of governments, by corporations, or individuals. It may or may not be legal and may or may not require authorization from a court or other independent agency.

Computer and network surveillance programs are widespread today, and almost all Internet traffic is or could potentially be monitored for clues to illegal activity.

C.6.5.3 End-to-End Encryption

End-to-end encryption is a digital communications paradigm of uninterrupted protection of data traveling between two communicating parties. It involves the originating party encrypting data so only the intended recipient can decrypt it, with no dependency on third parties. End-to-end encryption prevents intermediaries, such as Internet providers or application service providers, from discovering or tampering with communications. End-to-end encryption generally protects both confidentiality and integrity.

C.6.6 Views of the Network

Users and network administrators typically have different views of their networks. Users can share printers and some servers from a workgroup, which usually means they are in the same geographic location and are on the same LAN, whereas a Network Administrator is responsible to keep that network up and running. A community of interest has less of a connection of being in a local area, and should be thought of as a set of arbitrarily located users who share a set of servers, and possibly also communicate via peer-to-peer technologies.

Network administrators can see networks from both physical and logical perspectives. The physical perspective involves geographic locations, physical cabling and the network elements (e.g., routers, bridges and application layer gateways) that interconnect via the transmission media. Logical networks, called subnets in the TCP/IP architecture, map onto one or more transmission media. For example, a common practice in a campus of buildings is to make a set of LAN cables in each building appear to be a common subnet, using virtual LAN (VLAN) technology.

Both users and administrators are aware, to varying extents, of the trust and scope characteristics of a network. Again using TCP/IP architectural terminology, an intranet is a community of interest under private administration usually by an enterprise, and is only accessible by authorized users (e.g., employees). Intranets do not have to be connected to the Internet, but generally have a limited connection. An extranet is an extension of an intranet that allows secure communications to users outside of the intranet (e.g., business partners, customers).

C.7 Network Structure

In general, every telecommunications network conceptually consists of three parts, or planes (so called because they can be thought of as being, and often are, separate overlay networks):

- The control plane carries control information (also known as signaling).
- The data plane or user plane or bearer plane carries the network's user traffic.
- The management plane carries the operations and administration traffic required for network management.

C.8 Communication System

The information exchanged devices – through a network, or other media – is governed by rules and conventions that can be set out in technical specifications called communication protocol standards. The nature of a communication, the actual data exchanged and any state-dependent behaviors, is defined by its specification. In digital computing systems, the rules can be expressed by algorithms and data structures. Expressing the algorithms in a portable programming language makes the protocol software operating system independent.

Operating systems usually contain of a set of cooperating processes that manipulate shared data to communicate with each other. This communication is governed by well-understood protocols, which can be embedded in the process code itself.

In contrast, because there is no common memory, communicating systems have to communicate with each other using a shared transmission medium. Transmission is not necessarily reliable, and individual systems may use different hardware and/or operating systems. To implement a networking protocol, the protocol software modules are interfaced with a framework implemented on the machine's operating system. This framework implements the networking functionality of the operating system. The best known frameworks are the TCP/IP model and the OSI model.

At the time the Internet was developed, layering had proven to be a successful design approach for both compiler and operating system design and, given the similarities between programming languages and communication protocols, layering was applied to the protocols as well. This gave rise to the concept of layered protocols, which nowadays forms the basis of protocol design.

Systems typically do not use a single protocol to handle a transmission. Instead they use a set of cooperating protocols, sometimes called a protocol family or protocol suite.

C.9 Object-Oriented Programming

This is a programming paradigm based on the concept of "objects", which are data structures that contain data, in the form of fields, often known as *attributes*; and code, in the form of procedures, often known as *methods*. A distinguishing feature of objects is that an object's procedures can access and often modify the data fields of the object with which they are associated (objects have a notion of "this"). In object-oriented programming, computer programs are designed by making them out of objects that interact with one another. There is significant diversity in object-oriented programming, but most popular languages are class-based, meaning that objects are instances of classes, which typically also determines their type.

Many of the most widely used programming languages are multi-paradigm programming languages that support object-oriented programming to a greater or lesser degree, typically in combination with imperative, procedural programming.

Object-oriented programming attempts to provide a model for programming based on objects. Object-oriented programming integrates code and data using the concept of an "object". An object is an abstract data type with the addition of polymorphism and inheritance. An object has both state (data) and behavior (code). Objects sometimes correspond to things found in the real world. For example, a graphics program may have objects such as "circle", "square" and "menu".

Object orientation uses encapsulation and information hiding. Object-orientation essentially merges abstract data types with structured programming and divides systems into modular objects that own their own data and are responsible for their own behavior. This feature is known as encapsulation. With encapsulation, the data for two objects are divided so that changes to one object cannot affect the other. Note that all this relies on the various languages being used appropriately, which, of course, is never certain.

The object-oriented approach encourages the programmer to place data where it is not directly accessible by the rest of the system. Instead, the data is accessed by calling specially written functions, called *methods*, which are bundled with the data. These act as the intermediaries for retrieving or modifying the data they control. The programming construct that combines data with a set of methods for accessing and managing that data is called an object.

Defining software as modular components that support inheritance is meant to make it easy both to re-use existing components and to extend components as needed by defining new subclasses with specialized behaviors. This goal of being easy to both maintain and reuse is known in the object-oriented paradigm as the "open/closed principle". A module is open if it supports extension (e.g., can easily modify behavior, add new properties, provide default values, etc.). A module is closed if it has a well-defined stable interface that all other modules must use and that limits the interaction and potential errors that can be introduced into one module by changes in another.

C.10 Programming Tool or Software Development Tool

This is a computer program that software developers use to create, debug, maintain, or otherwise support other programs and applications. The term usually refers to relatively simple programs, that can be combined together to accomplish a task, much as one might use multiple hand tools to fix a physical object. The ability to use a variety of tools productively is one hallmark of a skilled software engineer.

The most basic tools are a source code editor and a compiler or interpreter, which are used ubiquitously and continuously. Other tools are used more or less depending on the language, development methodology and individual engineer, and are often used for a discrete task, like a debugger or profiler. Tools may be discrete programs, executed separately – often from the command line – or may be parts of a single large program, called an integrated development environment. In many cases, particularly for simpler use, simple *ad hoc* techniques are used instead of a tool, such as print debugging instead of using a debugger, manual timing (of overall program or section of code) instead of a profiler or tracking bugs in a text file or spreadsheet instead of a bug tracking system.

Index

Printed and bound by CPI Group (UK) Ltd, Croydon, CR0 4YY

16/04/2025

14658556-0002